ON SOUND

AND

ATMOSPHERIC VIBRATIONS,

WITH THE

MATHEMATICAL ELEMENTS OF MUSIC

ON SOUND

AND

ATMOSPHERIC VIBRATIONS,

WITH THE

MATHEMATICAL ELEMENTS OF MUSIC.

Designed for the Use of Students of the University.

BY

GEORGE BIDDELL AIRY,

M.A. LL.D. D.C.L.

GYAN

Published by

Gyan Publishing House
5, Ansari Road
Daryaganj, New Delhi-110002
Phone: 011-47034999, 9811692060
E-mail: books@gyanbooks.com

Distribution Network
gyanbooks.com
India, USA, Canada, UK, Australia

ISBN: 978-93-6208-904-5 (HB)
First Published, 1868

2nd Impression 2024

Printed at: Gyan Press, Delhi.

ON SOUND AND ATMOSPHERIC VIBRATIONS, WITH THE MATHEMATICAL ELEMENTS OF MUSIC
Author: GEORGE BIDDELL AIRY

ON SOUND

AND

ATMOSPHERIC VIBRATIONS,

WITH THE

MATHEMATICAL ELEMENTS OF MUSIC.

ON SOUND

AND

ATMOSPHERIC VIBRATIONS,

WITH THE

MATHEMATICAL ELEMENTS OF MUSIC.

DESIGNED FOR THE USE OF STUDENTS

OF THE UNIVERSITY.

BY

GEORGE BIDDELL AIRY, M.A. LL.D. D.C.L.

FORMERLY FELLOW, NOW HONORARY FELLOW, OF TRINITY COLLEGE;
LATE LUCASIAN PROFESSOR OF MATHEMATICS, AFTERWARDS PLUMIAN PROFESSOR
OF ASTRONOMY AND EXPERIMENTAL PHILOSOPHY;
IN THE UNIVERSITY OF CAMBRIDGE.
ASTRONOMER ROYAL.

1871.

IN PUBLISHING THIS TREATISE

I DESIRE TO RECORD MY SENSE

OF

THE EMINENT SERVICES RENDERED TO THE

UNIVERSITY OF CAMBRIDGE

AND

THE MANY AND IMPORTANT ACTS OF KINDNESS

CONFERRED ON MYSELF

BY

MY LATE FRIEND

GEORGE PEACOCK, D.D.

AND TO STATE THAT

MORE THAN THIRTY YEARS AGO

BY HIM I WAS URGED

TO WRITE A TRACT ON SOUND

FOR USE IN THE UNIVERSITY.

G. B. AIRY.

1868, MARCH.

ADVERTISEMENT

TO THE SECOND EDITION.

In this Edition, alterations have been made in several paragraphs, with the view of removing obscurities in the statements or reasonings. The only additions made are contained in Articles 54*, 54**, 54***, in the latter parts of Articles 64 and 65, and in Article 93**. The original numeration of the Articles in the First Edition is preserved unaltered.

CONTENTS.

SECTION I.

GENERAL RECOGNITION OF THE AIR AS THE MEDIUM WHICH CONVEYS SOUND.

Articles 1—2. Pages 1—3.

SECTION II.

PROPERTIES OF AIR, ON WHICH THE FORMATION AND TRANSMISSION OF SOUND DEPEND.

Articles 3—17. Pages 4—20.

SECTION III.

THEORY OF UNDULATIONS, AS APPLIED TO SOUND; AND INVESTIGATION OF THE PASSAGE OF A WAVE OF AIR THROUGH A CYLINDRICAL PIPE, OR OF A PLANE WAVE THROUGH THE ATMOSPHERE GENERALLY.

Articles 18—34. Pages 22—55.

SECTION IV.

INVESTIGATION OF THE MOTION OF A WAVE OF AIR THROUGH THE ATMOSPHERE CONSIDERED AS OF TWO OR THREE DIMENSIONS.

Articles 34*—59. Pages 56—128.

SECTION VII.

ON MUSICAL SOUNDS, AND THE MANNER OF PRODUCING THEM.

Articles 71—86. Pages 154—204.

SECTION VIII.

ON THE ELEMENTS OF MUSICAL HARMONY AND MELODY, AND OF SIMPLE MUSICAL COMPOSITION.

Articles 87—98. Pages 205—241.

SECTION IX.

ON INSTRUMENTAL MUSIC, AND THE ADAPTATIONS OF MUSIC REQUIRED BY SPECIAL INSTRUMENTS.

Articles 99—101. Pages 242—254.

SECTION X.

ON THE HUMAN ORGANS OF SPEECH AND HEARING.

Articles 102—106. Pages 255—268.

ERRATA.

Page 93, line 2, *for* circnlar *read* circular.

„ 111, line 7,

For

The velocity parallel to xy in the direction of radius

$$=\frac{x}{r}\cdot\frac{dX}{dt}+\frac{y}{r}\cdot\frac{dY}{dt}=2xy\cdot\frac{R}{r}+xy\cdot\frac{dR}{dr}=xy\left(\frac{2R}{r}+\frac{dR}{dr}\right).$$

Read

The velocity parallel to xy in the direction of radius

$$=\frac{x}{r}\cdot\frac{dX}{dt}+\frac{y}{r}\cdot\frac{dY}{dt}=2xy\cdot\frac{R}{r}+xy\cdot\frac{x^2+y^2}{r^2}\cdot\frac{dR}{dr}=xy\left(\frac{2R}{r}+\frac{x^2+y^2}{r^2}\cdot\frac{dR}{dr}\right).$$

Page 117, line 9 from bottom, *for* considerations *read* consideration.

„ 230, line 3 from bottom, *for* Adestes read Adeste.

ON SOUND

AND

ATMOSPHERIC VIBRATIONS,

WITH THE

MATHEMATICAL ELEMENTS OF MUSIC.

SECTION I.

GENERAL RECOGNITION OF THE AIR AS THE MEDIUM WHICH CONVEYS SOUND.

1. *Ordinary experience on the transmitting power of Air.*

We cannot, in strictness, say that air is the only medium which *can* convey sound. There are instances, for example, of persons who are totally deaf to sounds produced by excitement of the air, but who can hear the sound of a watch, or a bell, when held by the teeth; the sound being then undoubtedly conveyed by the bony and other portions of the head to the auditory nerves. And we shall hereafter refer to experiments in which the transmission of sound from considerable distances is obviously produced exclusively by fluid or solid bodies. Still it is evident, as matter of ordinary experience, that sound is in all familiar cases transmitted by air, and is in a great number of them produced by air. Thus,

The interposition of a pile of buildings, &c. diminishes the intensity of a sound coming from a distance; the partial closing of a window or door diminishes the intensity of a sound coming from without: the more effectual closing interrupts the sound entirely.

The sound of a bell within the receiver of an air-pump, struck by self-acting mechanism, is gradually diminished as the air is gradually extracted from the receiver.

In all shapes of the trumpet, the flute, and the organ, the sound is obviously produced by action on the air.

In firing a gun, the sound is produced by the sudden creation of a gas similar in its mechanical properties to air.

The notoriety of these and similar instances is sufficient to induce us to refer to the properties of air in our investigations of the Theory of Sound. It may however be remarked here that the theory of the transmission of sound through fluid and solid bodies will be easily connected with the simplest case of the transmission through air.

2. *Ordinary experience on diminution and retardation of Sound by distance.*

The first of the obvious laws of Sound in general is, that it diminishes with the distance. The accurate law of diminution will be considered hereafter when we have applied mathematical investigation to the theory. The second law, which is less obvious, but which is sufficiently well known, and has been remarked by observant persons in all ages (see, for instance, Lucretius, VI. 169, &c.) is, that the propagation of sound to a distance occupies time, and that the time required is sensibly proportional to the distance to be traversed. It is also well known that sounds of different pitch and of different loudness travel with sensibly the same speed: the sounds of a ring of bells, at whatever distance they are heard, fall on the ear in the same order. The velocity may be stated roughly to lie between 1000 feet and 1200 feet per second. The numbers, and their variation under certain circumstances, will be given with greater accuracy when we treat of the theoretical investigation.

SECTION II.

PROPERTIES OF AIR, ON WHICH THE FORMATION AND TRANSMISSION OF SOUND DEPEND.

3. *Equality of Pressure of Air in all directions.*

It is assumed, and we have no reason to doubt the law, that Air, as a fluid, is subject to the law of Equality of Pressure in all directions. If, for instance, air is forcibly confined in a vessel of glass or other material furnished with closed holes of equal dimensions on different sides, the confined air will exert equal pressures on the stoppers of those holes; and, if one of them is removed, the air will rush out with the same velocity, whichever be the hole selected, and whether the outburst be upwards, downwards, or in a horizontal or any other direction. It is also assumed that this fundamental property of each small volume of air holds when that small volume of air is in motion; although that motion might in some degree derange the laws of pressure in experimental cases like that to which we have alluded. Thus, if the pressure of air within a bottle were produced by the sudden rush of a quantity of air into the neck of the bottle, that part of the shoulder of the bottle near to its neck might not sustain the same pressure (on equal portions of surface) as the base of the bottle, because the inertia of the moving portions of air would produce the largest pressure on that part of the bottle whose resistance brought the air to a state of rest; although the Law of Equal Pressure applied to every small volume of air in motion.

Indeed, as all our experiments on air are made on air moving through space with great rapidity, it is impossible to deny the application of experimental results derived from air apparently at rest to air which is really in motion.

4. *Absence of friction among the particles of Air.*

It is also assumed, in the following investigations, that the particles or very small volumes of air can move among each other with perfect freedom from friction or viscosity. It would seem probable that the effect of such friction, &c. would be, not to alter materially the laws of vibration at which we shall arrive, but to produce the rapid extinction of motion.

5. *Statement of the three Laws affecting the Pressure of Air.*

But the properties of Air to which it is most important to call attention here are the Laws of Pressure of Air considered as an Elastic Gas. These are three, (I) the law connecting the elastic force of air (or, which is necessarily the same thing, the external pressure that compels the air to occupy only a certain limited space) with the density of the air, at a definite temperature; (II) the variation produced in that law by a permanent change of the temperature of the air; (III), the variation produced in that law by a sudden change of the compression of the air.

6. *Construction of the Barometer.*

In order to explain the experimental investigations upon which law (I) is established, it is necessary to describe in its essential points the common barometer. Take a straight tube of glass, not less than 32 inches long, open at one end and closed at the other, hold it for a short time with the closed end downwards, and pour quicksilver (mercury) into it till it is quite full: then carefully stop the open end, either by pressing it with the finger or by inserting any tight plug which can be easily withdrawn; then invert the tube so that its open end is downwards, dip that open end (before the plug is withdrawn) into a cistern of mercury, and, when the end is securely lodged below the surface of the mercury, withdraw the finger or plug. (Figure 1 represents the apparatus thus arranged, in a shape convenient for experiment: the glass tube being carried by a tripod stand which prevents the open end of the glass tube from touching the bottom of the cistern.) Immediately the surface of the mercury in the tube will fall, till its height above the surface of the mercury in the cistern is a quantity not absolutely constant but (when the place of experiment is near the level of the sea) seldom less than 28 English inches, and seldom greater than 31 inches. The space above the mercury, in the tube, is the most perfect vacuum that we can make. If the tube be inclined, so as to bring its upper end below the horizontal plane indicated by the surface of the mercury, the mercury will rise and will fill the tube to its top. This is the form known as the "Cistern Barometer;" with proper arrangements for maintaining the surface of the mercury in the cistern at a constant height, it is considered to be the best for standard barometers. Another form of the instrument is that represented in Figure 2, to which that of Figure 3 is sometimes preferred because the lower surface of the mercury can thus be brought vertically below the upper surface; these forms are known as the "Siphon Barometer," and they are the most convenient for portable barometers. For preparing these, the tube is placed nearly horizontal with the open leg upwards; mercury is poured into it and shaken into the closed leg, till the closed leg is filled; then the closed end is raised, and the mercury sinks in it till its upper surface has the same elevation above the surface of the mercury in the lower leg, which, in the Cistern Barometer, the surface of mercury in the tube has above that in the cistern. In practice, it is necessary in preparing either form of the barometer to make the tube very hot, even hot enough to boil the mercury, in order to expel aqueous vapour; and, in measuring the height of the surface of the mercury, it is necessary to take into account that depression known as "capillary depression," which depends on the diameter of the tube, being greatest in tubes of smallest diameter. It is also necessary to reduce the measure of the mercurial column to what it would have been at a definite temperature of the mercury; the temperature universally adopted is that of freezing water (32° of Fahrenheit, 0° of Reaumur or Centigrade).

7. *Measure of Atmospheric Pressure.*

When these arrangements are properly carried out, we have a measurable elevation of mercury in the tube, which can be explained in no way but by a pressure of air upon the surface of the mercury in the cistern or in the short tube, and which gives the means of measuring that pressure. The free horizontal surface of the mercury in the cistern (conceived as extended horizontally through the vertical column of mercury which rises from it into the tube) can only be kept at rest by the equality of pressures of the various columns which stand upon it; the greater number of these columns consist of atmospheric air, but one of them is "that part of the mercury in the long tube, which is above the horizontal plane of the free surface of mercury in the cistern or short tube," and this part can be measured. The pressure of the lofty air-atmosphere upon the cistern, in fact, must be the same as the pressure of a low mercury-atmosphere whose summit is no higher than the mercury in the barometer-tube. It is often convenient to refer to a definite elevation of the barometer: that adopted in English documents is usually

30 English inches; that adopted by Continental scientific men is usually 0·76 metre, or 29·9218 inches (the metre being 39·37079 inches). In the latter case, the pressure of the atmosphere upon a square inch of the mercury in the cistern is the weight of 29·9218 cubic inches of mercury. A little familiarity with metrical measures will perhaps be gained by making a calculation through the metrical system. Thus, the pressure upon a square centimetre of the mercury-surface will be the weight of 76 cubic centimetres of mercury. Drs Moll and Van Beek (*Philosophical Transactions*, 1824) have adopted as the weight of one cubic centimetre of mercury 13·5962 grammes, and the gramme (the weight of a cubic centimetre of water) is 15·4324 English grains, of which 7000 make a pound avoirdupois. So that the pressure upon a square centimetre in pounds avoirdupois will be

$$\frac{76 \times 13{\cdot}5962 \times 15{\cdot}4324}{7000}.$$

But as the centimetre is 0·393708 inch, the square centimetre is $(0{\cdot}393708)^2$ square inch. Thus, finally, we have for the pressure in pounds avoirdupois upon a square inch

$$\frac{76 \times 13{\cdot}5962 \times 15{\cdot}4324}{7000 \times (0{\cdot}393708)^2}.$$

The calculation by logarithms is easy, and the result is 14·6966 lbs. This is the pressure upon a square inch when the barometer stands at 29·9218 inches; when the height of the barometer is different, the pressure is altered in the same proportion.

8. *Height of Homogeneous Atmosphere, dry or humid.*

The authors cited above have given the weight of one cubic centimetre of dry air (found by exhausting the air from a bottle, and weighing the bottle empty of air and full of air) as 0·00129954 gramme, the air having been weighed when the height of the barometer was 0·76 metre, and the temperature of the air 0° Centigrade; and this gives us the means of determining the value of one very important constant. The weight of one cubic centimetre of mercury was found (see last page) to be 13·5962 grammes; and therefore the weight of mercury is $\frac{13 \cdot 5962}{0 \cdot 00129954}$ × the weight of air such as we have at the earth's surface under these circumstances; and therefore the pressure of air (with the barometric height and temperature mentioned above) is equal to the weight of a sea of air whose depth is

$$0 \cdot 76 \text{ metre} \times \frac{13 \cdot 5962}{0 \cdot 00129954},$$

provided that air were like water, an incompressible and homogeneous fluid; the weight being considered as produced by the action of gravity at that place (Paris) at which the experiments were made. This is usually called "the height of a homogeneous atmosphere." It will enter as a constant into every part of the investigations which follow. Taken in connexion with the value of the gravity under whose action it is estimated, it represents a fundamental element in the constitution of air. The numerical value formed from the numbers above is 7951·36 metres or 26087·6 English feet. We shall always use the symbol H for this element. It will appear shortly that the height of the homogeneous atmosphere is not invariable, but that it depends on the temperature of the air: for that variable height we shall use the symbol H'.

In this calculation we have omitted the consideration of the moisture in the atmosphere. It is to be understood that we possess means in ordinary use for ascertaining the amount of moisture in the air (usually based on the principle of observing how much the air must be cooled in order to make it deposit dew), and that, having examined the properties of vapour at various temperatures nearly as we have examined those of air, we know what is the elastic force of the vapour in the air. Also that we know the weight of vapour which exercises a given elastic force, and that it is about $\frac{5}{8}$ of the weight of dry air which exercises the same force. With this we must make use of "Dalton's Law," based on experiments which shew that, when dry air and vapour (or any other gases) are inclosed together in the same space, the elastic force which the mixture exerts is the sum of the elastic forces due separately to each. Suppose then that, with barometer and thermometer as above mentioned, we find that the elastic force of the vapour is $\frac{1}{n}$ of that of the dry air: (n in these countries is seldom less than 30). We now have a mixture of dry air producing a pressure, on the square centimetre, of

$\frac{n}{n+1} \times 0·76 \times 13·5962$ grammes, and of vapour producing a pressure of $\frac{1}{n+1} \times 0·76 \times 13·5962$ grammes: total $0·76 \times 13·5962$ grammes. And the weight of a cubic centimetre of this mixture is $\frac{n}{n+1} \times 0·00129954$ grammes $+ \frac{1}{n+1} \times \frac{5}{8} \times 0·00129954$ grammes: total

$$0·00129954 \text{ grammes} \times \left\{1 - \frac{3}{8(n+1)}\right\}.$$

Hence, the height of the homogeneous atmosphere will be

$$0·76 \text{ metre} \times \frac{13·5962}{0·00129954} + \left\{1 - \frac{3}{8(n+1)}\right\}$$

$$= 7951·36 \text{ metres} \times \frac{8n+8}{8n+5}.$$

When $n = 30$, the fraction increases the height of the homogeneous atmosphere by $\frac{1}{82}$ part (which, as we shall shortly see, increases the velocity of sound by $\frac{1}{164}$ part). We shall usually omit all mention of this.

9. *Measure of Elastic Force of Air under different circumstances.*

We are now in a state to consider the investigation of law (I) regarding the relation between the elastic force of air and the space which it occupies. We premise that in a glass tube, though we cannot everywhere measure the section, we can with great accuracy measure the capacity of the tube from a closed end to various points of the tube, by successively pouring in small quantities of mercury whose weights are known. Suppose then that in Figures 4 and 5 we have tubes of the siphon form, each containing air above the mercury in the closed leg. The quantity of that inclosed air will be known with accuracy by varying the quantity of mercury in the open leg till the mercury stands at the same height in the two legs: for then it is evident that the pressure of air on every unit of surface of the mercury in the closed leg is equal to the pressure similarly measured in the open leg, and the elastic state of the air in the closed leg is the same as that of the open air. Suppose now, that to produce the state of things in Figure 4, some quicksilver is withdrawn, or that to produce the state of things in Figure 5 some quicksilver is added. Then the equilibrium at the lower of the two surfaces is thus to be estimated. In Figure 4 a unit of the surface in the closed tube is pressed down by the elastic force of the inclosed air: to this is to be added the weight of the column of mercury whose height is the excess of height in the closed leg above height in the open leg; and thus is found the pressure upon a unit in the horizontal section of the mercury in the closed leg at the height of the surface in the open leg. This must be balanced by the pressure upon a unit in the surface in the open leg; which pressure is merely the atmospheric pressure; that is, it is the pressure of a column of mercury whose height is the length of the barometric column found by the operations in Article 7. Hence the pressure of the inclosed air upon a unit of surface *plus* the pressure of the column of mercury whose length is the difference of heights of the two surfaces of mercury must equal the pressure of the barometric column. In the case of Figure 5, it will be found in the same way that the pressure of the inclosed air is equal to the pressure of the column whose length is the difference of heights of the two surfaces *plus* the pressure of the barometric column. By these operations we have obtained measures of the volume occupied by a given quantity of air, and of the corresponding pressure upon a unit of surface estimated by the height of a column of mercury.

10. *With given temperature: Elastic Force of Air is proportional to its Density.*

These experiments have been made frequently, and to great extents of compression and of expansion of the air inclosed in the tube. And the result, or Law (I) is the very simple one—"The pressure which a given quantity of air produces on a unit of surface is inversely proportional to the space occupied by that air." Or, since the diminution or increase of the space occupied necessarily increases or diminishes in the same proportion the density of the given quantity of air occupying that space, the Law (I) may be stated, "The pressure which air exerts upon a unit of surface is proportional to the density of the air." This is commonly cited as Boyle's Law, or Mariotte's Law. It supposes that the temperature of the air is the same in all the experiments.

11. *The Height of homogeneous Atmosphere, with given Temperature, is independent of its Density.*

One consequence of Law (I) is, that the element which we have called "the height of a homogeneous atmosphere" is independent of the density of the air. For, the element in question is that height of a column of air, of the same density, whose weight would produce the observed pressure; but, by this Law (I), when the density is increased or diminished, the observed pressure is found to be increased or diminished in the same proportion; and therefore the height of a column of air of this altered density, whose weight will produce this altered pressure, will be the same as before.

12. *Symbols, and Units of Measure.*

It may now be convenient to introduce symbols. Let D be the density of air under some normal circumstances. By D we mean the mass of the air contained in a cubic unit; the mass being measured by its equality with multiples of the unit of mass described as a weight, as ascertained by weighing. (Though gravity enters into the operation of weighing, its power or change of power affects the two subjects equally, and therefore this definition is independent of the measure of gravity at the locality of any experiment.) The unit of weight may be the grain, the pound, the gramme, &c., and the unit of measure may be the foot, the inch, the centimetre, &c. And let P be the pressure of air under the same circumstances. By P we mean the pressure which the air exerts upon a unit of surface, estimated not as a mass but as a weight; the weight being defined by the number of units of weight. (This definition does depend on the measure of gravity at the locality of the experiment; the greater is the gravity, the smaller will be the number of units of weight required to produce the observed pressure.) Now a column of the height H (expressed by the number of units of length) will contain H cubic units, each of which has the mass D and is weighed at the locality as D; and the whole column will weigh HD. This, when H means the height of a homogeneous atmosphere, is supported by the pressure which is measured by weight P. Therefore $P = H . D$.

13. *Algebraical expression for Pressure in terms of Density.*

If the space occupied by the air is changed without changing its temperature, and if P becomes Π, and D becomes Δ, then Law (I) asserts that $\frac{\Pi}{P} = \frac{\Delta}{D}$. Combining this equation with the last,

$$\Pi = H . \Delta.$$

14. *Dependence of Elastic Force of Air and of Height of Homogeneous Atmosphere on Temperature.*

We may now proceed with Law (II); but it requires no details, because the experiments are precisely the same in form as those for Law (I), the only difference being that the air is used at various temperatures, and the inferences from the various experiments are compared. And the result may be stated thus with sufficient accuracy. The effect of increasing the temperature of air is, to increase its elastic force if its volume is not altered, or to increase its volume if the compressive force answering to its elastic force is not altered; and the law, as depending on temperature, is, that the elastic pressure (in one case) or the volume (in the other case) may be represented by 450 + the degrees of Fahrenheit's scale. Thus the pressure with given quantity in a given space at the temperature 32° Fahrenheit will be to that with the same quantity in the same space at 50° Fahrenheit as 482 to 500. It is easily seen from this that if the symbol H be confined to the meaning "height of homogeneous atmosphere at the freezing point of water," then the corresponding height H', for the temperature 50° Fahrenheit, will be $\frac{500}{482} \times H$.

15. *Rise or Fall of Atmospheric Temperature, produced by sudden Contraction or Expansion.*

Law (III) applies to a very remarkable property of air, which is not recognized, we believe, as affecting any other theories of natural philosophy connected with the atmosphere, but which is of the utmost importance in relation to the theory of Sound. In the experiments described in Articles 9 and 10 the operations are not rapid, and great pains have been taken to make the temperatures perfectly uniform, through the changes of pressure in each experiment. And the Law (I), or Boyle's or Mariotte's Law, holds true only on the supposition that the temperature of the air is the same with air much compressed and with air little compressed. But when the changes of volume and pressure are very rapid, the changes of temperature of the air are very great. Upon suddenly condensing air it becomes very hot. We have verified the experiment that; if inflammable tinder is placed in the bottom of a cylinder in which a piston fits tightly and slides easily; when the piston is driven rapidly down so as to condense the air very much before it has had time to impart the whole of its caloric to the surrounding metal, the air will inflame the tinder. And we have remarked, in the powerful air-pumps (driven by large steam-engines) which were used to exhaust the air-tubes upon the Atmospheric Railway, that when the attenuated air in the tube, having acquired the temperature of the ground, was compressed by the operation of pumping so as to be able to open the last valve in opposition to the pressure of atmospheric air, the emergent air was so hot as to be unbearable to the hand. If the heated air, without having lost caloric, be allowed to expand to its former dimensions, it exhibits its former temperature: that is, it cools by sudden expansion. And this is so well known that it has been proposed to supply apartments in hot climates with cool air, by compressing air in a close vessel, allowing the increased heat to escape by contact of the vessel with the external air or neighbouring substances, and then permitting the condensed air (at the atmospheric temperature) to expand into the apartments, when it would have a much lower temperature.

16. *Alteration of the law connecting Elastic Force or Pressure with Density, by the circumstance last mentioned.*

It follows that when the changes of volume of the air are rapid (and in the theory of Sound we shall have to treat of changes which are never so slow as 30 in a second of time, and sometimes as quick as 4000 in a second), the equation $\Pi = P.\frac{\Delta}{D}$ cannot hold. For, suppose that the air is suddenly compressed, or that Δ is greater than D, then the heat is increased above that which is supposed in the equation; the elasticity is increased (by Law (II)); and $\Pi > P\frac{\Delta}{D}$. On the contrary, when the air is suddenly expanded it is cooled; the heat and elasticity are less than the equation contemplates; and $\Pi < P\frac{\Delta}{D}$. In both cases the pressure may be represented, at least approximately, by the formula $\Pi = P\left(\frac{\Delta}{D}\right)^N$, where N exceeds 1. It is not easy to ascertain the value of N from experiment. Some physicists have endeavoured to infer it from considerations of the commutability of caloric and *vis viva.* The reader will find much information on these points in the *Philosophical Magazine*, 1844, 5, 7, 8, 9, 1851, and later years; *Philosophical Transactions*, 1824 and 1830; *Mécanique Céleste,* Vol. v.; *Journal de Physique, &c.* Different values assigned for N are 1·333; 1·348; 1·3748; 1·421; 1·4254; 1·4954. We are inclined to adopt $\frac{36}{25}$ or 1·44 as not far from the truth. If for $\sqrt{N}$ (which we shall often use) we put the symbol n, then $n = \frac{6}{5} = 1{\cdot}2$. We know not whether this is varied by variation in the original temperature of the air.

17. *Collection of the Laws affecting the pressure of Air.*

Thus, on combining the different laws, we find these values for the elastic pressure of air measured as is stated in Article 12.

When the air, which is the subject of experiment, is allowed to assume the temperature expressed by the thermometer-reading for surrounding objects,

$$\Pi = H'.\Delta$$

$$= H \times \frac{450 + \text{reading of Fahrenheit's thermometer}}{482} \times \Delta.$$

When the change in the state of the air is very rapid,

$$\Pi = P.\left(\frac{\Delta}{D}\right)^N = H'.D.\left(\frac{\Delta}{D}\right)^N$$

$$= H \times \frac{450 + \text{reading of Fahrenheit's thermometer}}{482} \times \frac{\Delta^N}{D^{N-1}},$$

where we have reason to think that N or n^2 does not differ greatly from 1·44, but where we have no knowledge as to the possible dependence of the value of N on the thermometer-reading.

It will be convenient hereafter to put Θ or θ^2 for the fraction expressing the known thermometrical factor.

Section III.

THEORY OF UNDULATIONS, AS APPLIED TO SOUND; AND INVESTIGATION OF THE PASSAGE OF A WAVE OF AIR THROUGH A CYLINDRICAL PIPE, OR OF A PLANE WAVE THROUGH THE ATMOSPHERE GENERALLY.

18. *General conception of a Wave.*

The theory of the transmission of Sound through the air (as well as through other bodies) is essentially founded upon the conception of the transmission of waves, in which the nature of the motion is such, that the movement of every particle is limited, while the law of relative movement of neighbouring particles is transmitted to an unlimited distance, either without change or with change following a definite law. For better understanding of this conception, the reader is referred to Figure 6. The upper line (α) is intended to represent the position of particles of air, *a*, *b*, *c*, *d*, &c. at uniform distances, in the state of quiescence; the next line (β) represents them at a certain time T in a different state, in which they have been placed by some artificial cause, more closely condensed about *a*, about *a′*, &c.; and more widely expanded about *g*, about *g′*, &c.; the third line (γ) represents them at the time $T+\frac{\tau}{4}$ in a state analogous to the second, but with the points of condensation about *d*, *d'*, &c., and the points of expansion about *k*, *k'*, &c.; the fourth line (δ) shews the state of condensation and rarefaction as having travelled still further in the same direction; and so on for the successive lines ϵ, ζ. Now if the places of the points in these lines represent the positions of the particles at successive equal intervals of time, it is plain that we have states of condensation and states of rarefaction travelling on continually without limit, in one direction; while the motion of every individual particle is extremely small, and is alternately backwards and forwards. And this is the conception of a wave as depending on the motion of particles in the same line as that in which the wave travels; this is the kind of wave which we shall consider as explaining the transmission of Sound.

But there are other kinds of movements of particles, which are equally included under the conception of wave. For instance, in Figure 7, the motion of the particles is entirely transverse to the horizontal lines of the diagram; and, here, it is not states of condensation and rarefaction that travel continually in the same direction, but states of elevation and depression that so travel. (This is the kind of wave which is recognized as applying to polarized light.) In Figure 8, the motion of the particles consists of a combination of the two motions in Figure 6 and Figure 7; the vertical displacement of the particles so accompanying the horizontal displacements, that the places where the particles are most condensed in the horizontal direction are the places where they are most elevated in the vertical direction. (This is the character of waves of water.)

But in all these there is one general character; that a *state of displacement* travels on continually in one direction, without limit; while the motion of each individual particle is or may be small and of oscillatory character. And this is the general conception of a wave. It will be remembered that the special character of the waves of air applying to the problem of Sound is, that the displacements of the particles are in the same direction (backwards and forwards) as that in which the wave travels.

19. *The idea of a Wave was first entertained and developed by Newton.*

This idea appears to have been first entertained by Newton, and was certainly first developed by him, for the purpose of explaining what till then was totally obscure, the transmission of Sound through Air; it is worked out in the third book of the *Principia*, and among the many wonderful novelties of that wonderful work, it is not the least interesting or the least important. The mere conception of the motion of particles in the way pointed out above is a very small part of Newton's work; the really important step is, to shew that the condensations and rarefactions produced by these motions will, by virtue of the known properties of air, produce such mechanical pressures upon every separate particle that the different changes of motion, which those pressures will produce on each individual particle, will be such that the assumed laws of movement will necessarily be maintained. Perhaps it is not easy for us now to realize the boldness of the conception and the difficulties of the problem. It required a new Calculus, the Theory of Partial Differential Equations, of which this is the first instance. Newton himself could only solve the equation synthetically, not analytically; and, in consequence of the use of this method, he gave a restricted solution, not the general solution; but the solution contained no other error. To reconcile his theoretical inference for the Velocity of Sound with observed measures of velocity, he suggested the idea that the dimensions of the particles of air produced a sensible effect; we have in later times explained the discordance by the theory given above in Articles 15 and 16.

20. *Newton's treatment of Waves of Air.*

Newton's proposition 47 is headed, "When pulses are propagated through a fluid, every particle oscillates with a very small motion, and is accelerated and retarded by the same law as an oscillating pendulum," that is, by the law, X (the displacement of a particle) $= A\,.\,\text{cosine}\,(Bt - C)$. After a short explanation, he says, "Let us suppose then that the medium is by some cause put into such a state of motion, and let us see what follows." He then takes three particles at small intervals, and supposes that, in the circle where the cosine represents X, the points corresponding to these particles have different places; which amounts to the same as supposing that C is not constant but depending on x, or that $X = A$. cosine $(Bt - Dx)$. He then attaches each of the three values of X to the three original ordinates, thus forming the three disturbed ordinates $x + X$; he finds the space now between them; he finds the density and elasticity of the air between them, whose variable part it is easily seen is proportional to sine $(Bt - Dx)$; he then takes the excess of the front elasticity above the back elasticity, which evidently is proportional to cosine $(Bt - Dx)$: and as this backward force is proportional to X, he infers that these forces will account for a pendulum-like motion of the particles. Proposition 48 is in fact contained in Proposition 49, "Given the density and elastic force of the medium, to find the velocity of the pulses." He supposes a pendulum to be constructed whose length is the height of homogeneous atmosphere (our H). He then says, as the theorem to be proved, "In the time occupied by a complete or double oscillation of that pendulum, the pulse will pass over the space $2\pi H$." It is difficult to give an idea of the process without copying Newton's very words; but it depends on estimating, from the preceding considerations, the time of complete oscillation of a disturbed particle, or $\frac{2\pi}{B}$, and remarking that in that time the pulse must have passed over a space equal to the interval beween two waves, or $\frac{2\pi}{D}$. (This will be seen upon examination of Figure 6; in tracing the successive states of each particle *a, d, g,* &c. at each time $T, T+\frac{\tau}{4}, T+\frac{2\tau}{4}, T+\frac{3\tau}{4}, T+\tau$, it will be seen that every particle has gone through its complete oscillation backwards and forwards between the time T and the time $T+\tau$; and it will also be seen that, between the time T and the time $T+\tau$, the state of condensation has travelled forward with uniform velocity from a to a'.) And, having found this, he infers that the velocity will be that acquired by falling through $\frac{H}{2}$. The whole process is most ingenious and accurate; only deficient in generality, in supposing (apparently) that no other law of motion would satisfy the conditions. He then, with inaccurate weight of air, finds $H = 29725$ feet, and theoretical velocity of sound $= 979$ feet per second: to which he adds $\frac{1}{9}$ part for the supposed magnitude of particles of air. And he remarks that the velocity will increase with the temperature. He also endeavours to take account of the aqueous vapour in the air.

21. *Algebraical treatment of Waves of Air travelling along a tube; and formation of the Partial Differential Equation, neglecting small quantities.*

The algebraical method of treating the problem will be as follows. Let x be the distance (measured parallel to the axis of the tube) of any particle from the origin of measure of x, in the quiescent state, that is, in the state in which the density of the air in every part of the tube is represented by D; and at any time t let the particle be in a place advanced beyond its original place by the quantity X, so that its ordinate is now $x+X$; X is different for different particles at the same time, and is different for the same particle at different times, and therefore is a function of both x and t. Now consider the place of a particle whose primary distance from origin was $x+h$; its present distance from origin is

$$x+X+h+\frac{dX}{dx}h+\frac{d^2X}{dx^2}\cdot\frac{h^2}{1.2}+\&c.;$$

its present distance from the particle before mentioned is

$$h\left(1+\frac{dX}{dx}\right)+\frac{d^2X}{dx^2}\cdot\frac{h^2}{1.2}+\&c.;$$

the mass of air which, with density D, did occupy the length h, now with density Δ occupies the length

$$h\left(1+\frac{dX}{dx}\right)+\frac{d^2X}{dx^2}\cdot\frac{h^2}{1.2}+\&c.;$$

whence

$$Dh = \Delta h\left(1 + \frac{dX}{dx}\right) + \Delta \frac{d^2X}{dx^2} \cdot \frac{h^2}{1.2} + \&c.;$$

or, forming the expression for Δ and supposing h indefinitely small,

$$\Delta = \frac{D}{1 + \frac{dX}{dx}} = D\left\{1 - \frac{dX}{dx} + \left(\frac{dX}{dx}\right)^2 - \left(\frac{dX}{dx}\right)^3 + \&c.\right\}.$$

For the present, we shall suppose the relative movements of the particles to be so small, that the higher powers of $\frac{dX}{dx}$ may be neglected; then our equation becomes this;

About the particle whose original ordinate was x, the density of the air, or Δ, is represented by $D - D\frac{dX}{dx}$.

From this it follows, by the theorem of Article 17, that,

About the particle whose original ordinate was x, the elastic pressure of the air upon a unit of surface, estimated as in Article 12, is $H' \times \frac{\Delta^N}{D^{N-1}}$, or $H'.D \times \left(1 - \frac{dX}{dx}\right)^N$;

or, still neglecting the higher powers of $\frac{dX}{dx}$,

$$\Pi = H'.D \times \left(1 - N\frac{dX}{dx}\right).$$

The same formula, when applied to a particle whose original ordinate was $x+k$, gives this result,

About the particle whose original ordinate was $x+k$, the elastic pressure of the air upon a unit of surface is

$$\Pi' = H'.D - H'.D.N\frac{dX}{dx}$$
$$- H'.D.N\frac{d^2X}{dx^2}k - H'.D.N\frac{d^3X}{dx^3}.\frac{k^2}{1.2} - \&c.$$

The mass of air included between these two particles, taking a tube whose section is 1, is Dk; and the pressure urging it forward is $\Pi - \Pi'$ or

$$H'.D.N\frac{d^2X}{dx^2}k + H'.D.N\frac{d^3X}{dx^3}.\frac{k^2}{1.2} + \&c.$$

Hence, remarking that in Article 12 all our pressures are estimated by weights, we have for the motion of the included air*,

* When a body whose weight is W falls freely under the action of gravity, it is in fact a mass W (estimated in conformity with the rules of Article 12) whose motion is affected by a pressure W (estimated in conformity with the rules of the same article). In this instance, as we know, the increase of velocity downwards produced in the unit of time is g. Hence we have, *in this case*,

$$\text{Increase of velocity in the direction of the force, produced in the unit of time} = \frac{\text{Pressure}}{\text{Mass}} \times g.$$

Therefore as, by the understood laws of motion, velocity produced is as pressure directly and as mass inversely, we shall have *in every case*

$$\frac{d^2(x+X)}{dt^2}=\frac{g}{Dk}\left\{H'.D.N\frac{d^2X}{dx^2}k+H'.D.N\frac{d^3X}{dx^3}\cdot\frac{k^2}{1.2}+\&c.\right\}.$$

But x is independent of t, so that $\frac{d^2x}{dt^2}=0$. Taking the rest on the supposition that k is made indefinitely small,

$$\frac{d^2X}{dt^2}=NgH'.\frac{d^2X}{dx^2}.$$

Or, putting θ^2 for the thermometer-factor in Article 17, and n^2 for N,

$$\frac{d^2X}{dt^2}=n^2.\theta^2.gH.\frac{d^2X}{dx^2}.$$

$$\left.\begin{array}{c}\text{Increase of velocity in the direction of the}\\ \text{force, produced in the unit of time}\end{array}\right\}=\frac{\text{Pressure}}{\text{Mass}}\times g;$$

provided that the pressures and masses are estimated as in Article 12.

Now, if X be the variable ordinate in the direction of motion, $\frac{dX}{dt}$, or the limit of $\frac{\text{increase of ordinate}}{\text{increase of time}}$, is the velocity; and the limit of $\frac{\text{increase of velocity}}{\text{increase of time}}$ (which for such a force as gravity is the same as increase of velocity produced in the unit of time) is

$$\frac{d.\,\text{velocity}}{dt},\ \text{or}\ \frac{d^2X}{dt^2}.$$

Hence our equation becomes

$$\frac{d^2X}{dt^2}=\frac{\text{Pressure}}{\text{Mass}}\times g.$$

We shall often have occasion to use this equation.

22. *The equation is independent of Local Gravity.*

The factor gH is purely an atmospheric element. For it will be remarked in Article 12 that, in applying measures to a given state of air, D is independent of the gravity at the place of experiment, but P is inversely as the gravity; and therefore, as $P = H . D$, H is inversely as the gravity. Hence, wherever the experiments are made, gH is invariable. We shall put for it the symbol a^2. Now at Paris where the weights of air, &c. were determined, the length of the seconds pendulum $= 39{\cdot}12877$ inches (*Encyclopædia Metropolitana*, 'Figure of Earth,' section 8), whence $g = 32{\cdot}18212$ feet. In Article 8 we have found $H = 26087{\cdot}6$ feet. Using the English foot as the unit of length and the mean solar second as the unit of time,

$$a = \sqrt{gH} = 916{\cdot}2722.$$

Our partial differential equation now is

$$\frac{d^2X}{dt^2} = n^2 . \theta^2 . a^2 . \frac{d^2X}{dx^2}.$$

In the observed phænomena of sound we have very strong reason for believing that n does not depend on the nature of the motion of the particles of air; but we have no means of knowing how it may depend on the temperature of the air. In any case it may be combined with θ to form one factor. We shall in calculation consider n constant and equal to $1{\cdot}2$.

23. *General Solution of the Equation.*

The general solution of the equation above (see the Author's *Elementary Treatise on Partial Differential Equations*, Article 35, making $\alpha = 0$), is

$$X = \phi\,(n\theta a\,.\,t - x) + \psi\,(n\theta a\,.\,t + x),$$

where the forms of the functions ϕ and ψ are absolutely undetermined by the theory of the solution, and are to be determined so as to answer to the physical conditions which are to be satisfied. Thus the solution admits of infinite variety. If we suppose

$$X = m \times (n\theta a\,.\,t - x) + m \times (n\theta a\,.\,t + x),$$

$$\text{or} = 2mn\theta a\,.\,t,$$

we have simply a uniform current through the tube, with equal velocity for all the particles. If

$$X = -\,m \times (n\theta a\,.\,t - x) + m \times (n\theta a\,.\,t + x),$$

$$\text{or} = 2mx,$$

so that the original ordinate x is changed into $x + X$ or $x + 2mx$, we have the air in a quiescent state, with the original intervals of its particles multiplied by $1 + 2m$, denoting a uniformly increased or diminished density throughout the tube, and implying that the ends of the tube are stopped. With second or higher powers, we should have movements produced by variable densities. But, for our Theory of Sound, we shall most frequently treat each of the functions in a general form.

24. *One term of the solution indicates a Wave travelling Forwards; the velocity is independent of the character of the wave.*

So far as depends on the function $\phi(n\theta a . t - x)$, whatever may be the form of ϕ, the following property holds. Suppose t increased by t'; and consider the state (at that increased time) of a particle whose original ordinate was $x + n\theta a . t'$. In the function, for t substitute $t + t'$, and for x substitute $x + n\theta a . t'$. Then the value of X becomes $\phi(n\theta a . t + n\theta a . t' - x - n\theta a . t')$, or $\phi(n\theta a . t - x)$; which is exactly the same value as that for the particle x at the time t. That is, if we consider the motion of a point whose quiescent ordinate was x' or $x + n\theta a . t'$, we find that, at the end of the time $t + t'$, its displacement is exactly the same as was the displacement of a point whose quiescent ordinate was x, at the end of the time t only. That is, if we increase the time, we may find certain particles in the same state of disturbance as the first particles at the first time, but we must go to a larger value of x in order to find these disturbed particles. This is exactly the characteristic of a wave. And since it appears that, upon increasing the time by t', we must go to a value of x increased by $n\theta a . t'$, it follows that the velocity of the wave is $n\theta a$, or $n\theta \times 916{\cdot}2722$ feet per second.

It is important to observe that this result is entirely independent of the character of the wave. It may be a single wave, symmetrical in its beginning and ending; it may be unsymmetrical; it may be a series of short waves, similar or dissimilar; it may be a series of long waves; the expression for displacement may be multiplied by a coefficient; or there may be any other conceivable variation; in all cases the result for velocity is the same. We shall see hereafter (Articles 71, 72, and 87), that the pitch of sound depends on the frequency of waves, and therefore on their length; and the quality depends on the form of the function. Thus it appears that in theory (as was remarked experimentally in Article 2) sounds of different pitch and quality travel with the same velocity.

25. *The other term represents a Wave travelling Backwards.*

But the general solution contains also the function $\psi(n\theta a \,.\, t + x)$. Here, to reproduce the same value of the function, if we increase t by t', we must *diminish* x by $n\theta a \,.\, t'$. This evidently means that the term represents a wave whose motion is in the direction opposite to the measure of x. So that the complete solution of the differential equation represents two waves, moving in opposite directions, and coexisting; the complete value of the disturbance X being the algebraic sum of the disturbances corresponding separately to the two separate waves. In treating of the disturbance of air in musical pipes, we shall find it necessary hereafter to take into account the two waves as simultaneously existing; but for the present, in treating of the direct transmission of sound, we require only one function or wave, $X = \phi(n\theta a . t - x)$.

26. *The differential equation of Article 21 being linear, if any number of different solutions be found, the sum of these solutions will also be a solution: this indicates the possibility of coexistence of waves, each of which singly is possible: but the theorem does not apply when the equation contains higher powers of the differential coefficients.*

Suppose that, having the equation $\frac{d^2X}{dt^2} = c^2\frac{d^2X}{dx^2}$, we obtain the solutions $X = A$, $X = B$, $X = C$, &c., where A, B, C, &c. are explicit functions of t and x. This means that the following equations are true:

$$\frac{d^2A}{dt^2} = c^2\frac{d^2A}{dx^2},$$

$$\frac{d^2B}{dt^2} = c^2\frac{d^2B}{dx^2},$$

$$\frac{d^2C}{dt^2} = c^2\frac{d^2C}{dx^2},$$

&c.,

and therefore, adding all together,

$$\frac{d^2(A + B + C + \&c.)}{dt^2} = c^2\frac{d^2(A + B + C + \&c.)}{dx^2};$$

which is the same as the original equation, putting $A + B + C + \&c.$ in the place of X. Consequently $A + B + C + \&c.$ is a solution of the equation. But A, B, &c., in the instances of Articles 24 and 25, represent different waves. It appears therefore that we may have a combination of different waves, each of which might exist alone: the characteristic of the combination being this, that the displacement X or $A + B + C$ in the combination will be the algebraic sum of all the displacements in the separate waves. It is evident that this result is not confined to the equation of Article 21, but that it applies to all equations in which

$$X, \ \frac{dX}{dx}, \ \frac{dX}{dt}, \ \frac{d^2X}{dx^2}, \ \frac{d^2X}{dt^2}, \ \frac{d^2X}{dx\,dt}, \ \&c.,$$

enter only to the first power.

But if we had such an equation as

$$\frac{d^2X}{dt^2} = a\left(\frac{dX}{dx}\right)^2,$$

and if we had found solutions A, B, &c., so that

$$\frac{d^2A}{dt^2} = a\left(\frac{dA}{dx}\right)^2,$$

$$\frac{d^2B}{dt^2} = a\left(\frac{dB}{dx}\right)^2,$$

$$\&c.;$$

and if we add these together,

$$\frac{d^2(A + B + \&c.)}{dt^2} = a\left\{\left(\frac{dA}{dx}\right)^2 + \left(\frac{dB}{dx}\right)^2 + \&c.\right\}.$$

But this is *not* the same as the original equation, putting $A+B+$ &c. in the place of X; for that would have required

$$\frac{d^2(A+B+\&c.)}{dt^2}=a\left(\frac{dA}{dx}+\frac{dB}{dx}+\&c.\right)^2,$$

which is a different equation. Hence the conclusions above stated, regarding the sum of solutions and the combination of waves, do not hold here.

27. *Plane Wave in Air of Three Dimensions.*

If we have a great number of pipes side by side, with waves of similar character passing simultaneously through all, so that, measuring x from a plane which is normal to all the pipes, the value of X in every pipe is represented by the same form and same coefficients of the function $\phi\,(n\theta a\,.\,t-x)$, the collateral condensations and pressures of air in the adjacent pipes will be the same, and there will be no tendency of the air in one pipe to press sideways into another pipe. We may therefore remove the material boundaries of these pipes; and then we have air, extended in three dimensions, through which passes a wave whose front is a plane, that is, in which all the points of similar motion and similar density are always in one plane. We shall not here delay longer on this subject, as it will be hereafter treated, perhaps more conveniently, by the general process of the "characteristic function."

28. *Limitations on the form of the Functions in the Solution.*

We shall now allude to the limitations on the form of the functions ϕ and ψ. For convenience we shall frequently put u for $n\theta at + x$, and v for $n\theta at - x$: so that the expression for X is

$$\phi(v) + \psi(u).$$

Now in the *Partial Differential Equations*, Articles 23 and 32, it appears that, as regards the mere algebraical solution of the partial differential equation, the form of the functions is absolutely unlimited; they may be discontinuous in any way, without reference to any algebraical formula, and with any degree of suddenness of change in the numerical value of ϕ, ϕ', ϕ'', &c. But in the physical problem we are limited by the suppositions tacitly made in the investigation which produced the partial differential equation. First, then, confining ourselves to one function, there can be no numerical discontinuity in X or $\phi(v)$. For, in forming

$$\frac{dX}{dt}, \text{ or } \frac{dX}{dv}\cdot\frac{dv}{dt}, \text{ or } n\theta a\,.\frac{dX}{dv},$$

by the limit of $\frac{\delta X}{\delta v}$, a numerical discontinuity or sudden interruption in the value of X would make δX finite while δv is indefinitely diminished: and therefore

$\frac{dX}{dv}$ and $\frac{dX}{dt}$ would at that point be infinite, and our investigation in Article 21 would be entirely inapplicable. Secondly, there can be no numerical discontinuity in $\phi'(v)$. For, in forming $\frac{d^2X}{dt^2}$, or $\frac{d}{dt}\left(\frac{dX}{dt}\right)$, or $n\theta a.\frac{d}{dv}\left(n\theta a.\frac{dX}{dv}\right)$, by the limit of $n\theta a.\frac{\delta.\phi'(v)}{\delta v}$, a discontinuity or sudden interruption in the value of $\phi'(v)$ would in like manner render $\frac{d^2X}{dt^2}$ infinite at that point; that is to say, there would be infinite force, infinite condensation of air, &c., all which is opposed to the ideas under which the investigation of Article 21 has been carried on. If, however, $\phi(v)$ and $\phi'(v)$ are free from discontinuity, then the effect of a numerical discontinuity in $\phi''(v)$ would be that at special values of v the magnitude of the forces, condensations, &c., would change suddenly; but there does not appear to be any physical impossibility in this. If $\phi(v)$, $\phi'(v)$, $\phi''(v)$, are free from numerical discontinuity, then there is no sudden change even in the magnitude of the force or condensation. We conclude, however, that it is sufficient that the two first terms, $\phi(v)$ and $\phi'(v)$, be free from numerical discontinuity.

29. *Forms proper for the Functions representing Continuous Series of similar Waves.*

If we propose to represent a series of waves in which all the successive waves are exactly similar, the most general formula which we can adopt for $\phi(v)$ is a function of $\sin v$ and $\cos v$; and if we reject fractional and negative powers of $\sin v$ and $\cos v$ (which is necessary in order that $\phi(v)$, $\phi'(v)$, $\phi''(v)$, &c. may never be infinite, and that $\phi(v)$ may never be ambiguous), that is, if we adopt only integral powers of $\sin v$ and $\cos v$, we can always (see Article 74) give the following form to the function,

$$A_1 . \sin(Bv + C_1) + A_2 . \sin(2Bv + C_2) + \&c.$$

which satisfies the conditions of Article 28.

The phænomena of music will usually be referred to this series: and, in most instances, to the first term alone.

30. *Introduction of the terms 'length of wave,' 'period of wave,' 'frequency of wave;' relation between their values and that of the 'velocity of waves;' remarks on the 'amplitude of vibration,' and its independence of the other quantities.*

The displacement X of a particle being represented, in a continuous series of waves, by the expression

$$X = A_1 . \sin(Bv + C_1) + A_2 . \sin(2Bv + C_2) + \&c.$$

upon making this maximum with respect to v, and positive (or negative, only confining ourselves to one sign), we find a definite value V for v, defined by numerical values (in terms of A_1, A_2, &c., C_1, C_2, &c.)

of sin BV and cos BV. These correspond to only one value of BV less than 2π, but they correspond also to $BV \pm 2\pi$, $BV \pm 4\pi$, &c. And upon substituting in the expression for X, they all give the same value, which is every where the maximum. Thus we find that the maximum recurs, and the general character of the wave recurs, when Bv is increased or diminished by $2m\pi$, or when v is increased or diminished by $\frac{2m\pi}{B}$.

Now $v = n\theta a \, . \, t - x$. If then we confine our attention to the wave at a certain instant of time, that is, if we regard t as constant; and if we survey the long series of waves, that is, if we contemplate the displacements and motions at that certain instant, of different particles; we find that when x is increased or diminished by an integer multiple of $\frac{2\pi}{B}$ (and at no other places) we come to particles in the same state of disturbance as that which was first considered. It is plain therefore that the 'length of a wave' is $\frac{2\pi}{B}$.

But if we fix our attention on a certain particle, that is, if we regard x as constant; and if we examine its state of disturbance at different times, that is, if we consider different values of t; we find that the same state of disturbance recurs when $n\theta a \, . \, t$ is increased or diminished by an integer multiple of $\frac{2\pi}{B}$, or when t is increased or diminished by an integer multiple of $\frac{2\pi}{n\theta a \,.\, B}$. This quantity $\frac{2\pi}{n\theta a \,.\, B}$ is evidently the interval in time between the passage of two successive waves at the same point, and is therefore the 'period of wave.'

The 'frequency of wave,' or the number of waves that occur in the unit of time, is evidently $\frac{n\theta a \,.\, B}{2\pi}$.

The 'velocity of wave,' Article 24, is $n\theta a$. Therefore, comparing the expressions above, we find,

length of wave = period of wave × velocity of wave;

$$\text{or length of wave} = \frac{\text{velocity of wave}}{\text{frequency of wave}}.$$

Now in these expressions, the factors A_1 and A_2 &c. do not occur. If the formula for X is restricted to the first term $A_1 \sin (Bv + C_1)$, A_1 does not enter at all into the determination of V; if there are other terms, the quotients $\frac{A_2}{A_1}$, &c. enter, but not the absolute values of A_1, A_2, &c. Thus the theorems just found are independent of A_1, A_2, &c. But the 'amplitude of vibration,' or maximum range in the amount of X, does depend on the absolute values of A_1, A_2, &c.; when the formula for X contains only one term, the amplitude is $2A_1$. Thus it appears that the amplitude (on one hand) and the length, period, frequency, and velocity of wave (on the other hand) are perfectly independent.

31. *The Solitary Wave, and Functions proper to represent it; and Interpretation of their effect.*

It is important to examine the case of the Solitary Wave: a wave which has been created by a single disturbance that occupied a limited time and was followed by absolute quiescence. Its algebraical conditions will be the following:—

Till v has a certain value A, $\phi(v) = 0$.

When v has a value included between A and $A + B$, $\phi(v)$ is to have a real value.

When the value of v exceeds $A + B$, $\phi(v) = 0$.

The function $\phi(v)$ must be such that in no part there be numerical discontinuity in the values of $\phi(v)$ and of $\phi'(v)$.

When $v = A$, and also when $v = A + B$; $\phi(v)$ and $\phi'(v)$, depending on that form of the function which applies from $v = A$ to $v = A + B$, must $= 0$; inasmuch as the values of $\phi(v)$ and $\phi'(v)$, before $v = A$ and after $v = A + B$, are $= 0$; and numerical discontinuity is to be avoided.

Functions can be found, in infinite variety of form, which satisfy these conditions. For instance,

$$\phi(v), \text{ from } A \text{ to } A + B, = C \,.\, (v - A)^2 \,.\, (A + B - v)^2,$$

$$\phi(v), \text{ from } A \text{ to } A+B,$$

$$= D \, . \sin^2 \frac{\pi}{B}(v-A) = \frac{D}{2} \, . \text{versin} \, \frac{2\pi}{B}(v-A).$$

In the second form, Newton has expressly remarked that the wave may be solitary.

The functions just exhibited possess this property, that the value of X, beginning from 0 when $v = A$, becomes a real value, which increases, and again decreases, till it is again 0 when $v = A + B$; that is, the particle returns to its original place. But in some cases (as on the explosion of gunpowder) it is desirable to have a form of function which will shew that the particle, after undergoing the wave-disturbance, is left in a place more advanced than the original. Such functions as the following satisfy that condition, retaining also the other conditions of freedom from numerical discontinuity:—

$$C \, . \int_v (v-A)^2 \times (A+B-v)^2 \, ;$$

$$D \, . \int_v \sin^2 \frac{\pi}{B}(v-A) \, ;$$

the value of the integral commencing from $v = A$, and the function expressed by the integral being used for the value of X from $v = A$ to $v = A + B$, after which the value of X is to be constant, and is to be that given by the definite integral from $v = A$ to $v = A + B$.

If $z = -\dfrac{1}{v-A} - \dfrac{1}{A+B-v}$, the function ϵ^z and its integral satisfy the terminal equations to any order of differentials.

The precise form and extent of application of the functions must be determined from consideration of the initial circumstances. (See the *Partial Differential Equations*, Article 53, &c.) Suppose that the impulse which generates the wave is given to the particle where $x = 0$. For that particle, v, or $n\theta a \,.\, t - x$, is $= n\theta a \,.\, t$. Therefore, knowing the displacement of that particle for a sufficient number of values of t or of $n\theta a \,.\, t$, we can express it as a function ϕ (algebraical or merely numerical) of $n\theta a \,.\, t$, so that $X_0 = \phi\,(n\theta a \,.\, t)$. Then, at every other point, $X = \phi\,(n\theta a \,.\, t - x)$, with the same form of ϕ. The interpretation of this, subject to the conditions at the beginning of this Article, will best be given by examinations referring to two considerations.

First, what is the state of all the particles at a certain time τ? At that time, $v = n\theta a \,.\, \tau - x$, or $x = n\theta a \,.\, \tau - v$; where $v = A$, $x = n\theta a \,.\, \tau - A$; where $v = A + B$, $x = n\theta a \,.\, \tau - A - B$. Thus the conditions (beginning with the third) are these :—

For the particles where x is less than $n\theta a \,.\, \tau - A - B$, there is no displacement.

For the particles where x is greater than $n\theta a \,.\, \tau - A - B$ and less than $n\theta a \,.\, \tau - A$, there is displacement.

For the particles where x is greater than $n\theta a \,.\, \tau - A$, there is no displacement.

Second, what is the movement of the one particle whose original ordinate is ξ? For that particle,

$$v = n\theta a \,.\, t - \xi, \text{ or } t = \frac{v+\xi}{n\theta a}; \text{ when } v = A,\ t = \frac{A+\xi}{n\theta a};$$

when $v = A + B$, $t = \dfrac{A + B + \xi}{n\theta a}$.

Thus the conditions (beginning with the first) become these:—

Till the time $= \dfrac{A+\xi}{n\theta a}$, the particle is not displaced.

From the time $\dfrac{A+\xi}{n\theta a}$ to the time $\dfrac{A+B+\xi}{n\theta a}$, the particle is displaced.

After the time $\dfrac{A+B+\xi}{n\theta a}$, the particle is not displaced.

These two exhibitions give a complete account in an intelligible form of the meaning of the discontinuous function.

32. *Formation of the Equation when small quantities are not neglected, and approximate Solution.*

In Article 21, the investigation was completed by neglecting the powers of $\dfrac{dX}{dx}$ above the first. In the present Article, we shall treat of the solution when superior powers are taken into account. For the reasons mentioned in Article 26, we must confine ourselves to a single wave. The accurate solution is by no means easy. We would refer our readers to a paper by Mr Earnshaw, in the *Philosophical Transactions*, 1860, where the solution is exhibited by an elimination between two functions which cannot be effected in a general form. The process which we shall use here is the more cumbrous one of successive substitution.

The equation obtained in Article 21, retaining all powers, but omitting for convenience the symbols $n^2\theta^2$, and remembering that $gH = a^2$, is

$$\frac{d^2X}{dt^2} = a^2 \cdot \frac{d}{dx}\left\{\frac{dX}{dx} - \left(\frac{dX}{dx}\right)^2 + \left(\frac{dX}{dx}\right)^3 - \&c.\right\},$$

or $$\frac{d^2X}{dt^2} - a^2\frac{d^2X}{dx^2} = a^2 \cdot \frac{d}{dx}\left\{-\left(\frac{dX}{dx}\right)^2 + \left(\frac{dX}{dx}\right)^3 - \&c.\right\}.$$

To transform this into an equation in which the independent variables are $u = at + x$ and $v = at - x$, we shall use the process in the *Partial Differential Equations*, Article 35. This makes

$$\frac{d^2X}{dt^2} - a^2\frac{d^2X}{dx^2} = 4a^2 \cdot \frac{d^2X}{du \cdot dv}.$$

Also $\frac{dW}{dx}$, whatever W may be, is $= \frac{dW}{du} - \frac{dW}{dv}$. Thus the equation, divided by $4a^2$, becomes

$$\frac{d^2X}{du\,.\,dv}=\frac{1}{4}\left(\frac{d}{du}-\frac{d}{dv}\right)\left\{-\left(\frac{dX}{du}-\frac{dX}{dv}\right)^2\right.$$

$$\left.+\left(\frac{dX}{du}-\frac{dX}{dv}\right)^3+\&c.\right\}:$$

or, more conveniently,

$$\frac{d^2X}{dv\,.\,du}=\frac{1}{4}\left(\frac{d}{dv}-\frac{d}{du}\right)\left\{\left(\frac{dX}{dv}-\frac{dX}{du}\right)^2\right.$$

$$\left.+\left(\frac{dX}{dv}-\frac{dX}{du}\right)^3+\&c.\right\}.$$

We shall now proceed with the successive steps of solution.

First step. Neglect all the terms on the right hand. Then $\frac{d^2X}{dv\,.\,du}=0$; X (see *Partial Differential Equations*, Article 30) $=\phi(v)+\psi(u)$. As we propose to consider only a wave travelling in the direction of x increasing, Article 24 of this treatise, we shall neglect the second function, and adopt $X=\phi(v)$.

Second step. Substitute the value $\phi(v)$ for X in the first term on the right hand, and we have, since $\frac{d\,.\,\phi(v)}{du}=0$,

$$\frac{d^2X}{dv\,.\,du}=\frac{1}{4}\left(\frac{d}{dv}-\frac{d}{du}\right)\{\phi'(v)\}^2=\frac{1}{4}\frac{d}{dv}\{\phi'(v)\}^2.$$

Integrating with respect to v,

$$\frac{dX}{du} = \psi'(u) + \frac{1}{4}\{\phi'(v)\}^2.$$

Integrating with respect to u,

$$X = \phi(v) + \psi(u) + \frac{1}{4}u\{\phi'(v)\}^2.$$

As before, we shall neglect $\psi(u)$, and thus we have

$$X = \phi(v) + \frac{1}{4}u\{\phi'(v)\}^2.$$

Third step. Substitute the value just found for X in the first term on the right hand, and the value $\phi(v)$ in the second term. We shall have within the right-hand bracket the quantity $\left(\frac{dX}{dv} - \frac{dX}{du}\right)^2$,

or $$[\phi'(v) + \frac{1}{2}u \, . \, \phi'(v) \, . \, \phi''(v) - \frac{1}{4}\{\phi'(v)\}^2]^2,$$

or $$\{\phi'(v)\}^2 + u \, . \, \{\phi'(v)\}^2 . \, \phi''(v) - \frac{1}{2} \, . \, \{\phi'(v)\}^3.$$

Also we have $\left(\frac{dX}{dv} - \frac{dX}{du}\right)^3$ or $\{\phi'(v)\}^3$.

The sum, or the whole quantity within the bracket, is

$$\{\phi'(v)\}^2 + u \, . \, \{\phi'(v)\}^2 . \, \phi''(v) + \frac{1}{2}\{\phi'(v)\}^3.$$

Affecting this with the external operation

$$\frac{1}{4}\left(\frac{d}{dv}-\frac{d}{du}\right),$$

it becomes

$$\frac{d^2X}{dv\,.\,du}=\frac{1}{4}\cdot\frac{d}{dv}\{\phi'(v)\}^2-\frac{1}{4}\{\phi'(v)\}^2.\,\phi''(v)$$

$$+\frac{1}{8}\cdot\frac{d}{dv}\{\phi'(v)\}^3+\frac{1}{4}u.\frac{d}{dv}[\{\phi'(v)\}^2.\,\phi''(v)]$$

$$=\frac{1}{4}\cdot\frac{d}{dv}\{\phi'(v)\}^2+\frac{1}{24}\cdot\frac{d}{dv}\{\phi'(v)\}^3$$

$$+\frac{1}{4}u.\frac{d}{dv}[\{\phi'(v)\}^2.\,\phi''(v)].$$

Integrating with respect to v,

$$\frac{dX}{du}=\psi'(u)+\frac{1}{4}\{\phi'(v)\}^2+\frac{1}{24}\{\phi'(v)\}^3$$

$$+\frac{1}{4}u.\{\phi'(v)\}^2.\,\phi''(v).$$

Integrating with respect to u,

$$X=\phi(v)+\psi(u)+\frac{1}{4}u.\{\phi'(v)\}^2$$

$$+\frac{1}{24}u\{\phi'(v)\}^3+\frac{1}{8}u^2.\{\phi'(v)\}^2.\,\phi''(v).$$

Or, neglecting $\psi(u)$, and restoring t and x,

$$X = \phi(at-x) + \frac{at+x}{4}\{\phi'(at-x)\}^2$$
$$+ \frac{at+x}{24}\{\phi'(at-x)\}^3$$
$$+ \frac{(at+x)^2}{8}\{\phi'(at-x)\}^2 . \phi''(at-x).$$

When we are considering the state of a solitary wave at a great distance from the origin, v or $(at-x)$, which (see Article 31) is limited within the value B, may be considered very small with regard to x; and in the factors of the small terms, for $at+x$ or $2x+(at-x)$ we may put $2x$; and we have

$$X = \phi(at-x) + \frac{x}{2}\{\phi'(at-x)\}^2$$
$$+ \frac{x}{12}\{\phi'(at-x)\}^3$$
$$+ \frac{x^2}{2}\{\phi'(at-x)\}^2 . \phi''(at-x).$$

This degree of approximation will suffice for our present purposes.

33. *Progressive Change in the Character of the Wave.*

It is seen here that the law of displacement of the particles undergoes change as the wave travels on. The original function receives the addition of new functions, which are affected with multipliers depending on the distance of the disturbed point from the origin of the disturbance. Supposing that we assume the terms at which we have arrived in the last Article to suffice for our information of what happens when the wave has travelled through a considerable but not an enormous distance, we may interpret their effect thus:

If $\phi(v)$ be $\frac{D}{2}\left\{1-\cos\frac{2\pi}{B}(v-A)\right\}$, where for convenience we will put v' for $v-A$, or

$$\phi(v)=\frac{D}{2}\left(1-\cos\frac{2\pi v'}{B}\right);$$

then we have

$$\phi'(v)=\frac{\pi D}{B}.\sin\frac{2\pi v'}{B};$$

which is 0 when $v'=0$, is + from $v'=0$ to $v'=\frac{B}{2}$, is 0 when $v'=\frac{B}{2}$, is − from $v'=\frac{B}{2}$ to $v'=B$, and is 0 when $v'=B$.

$$\phi''(v)=\frac{2\pi^2 D}{B^2}\cos\frac{2\pi v'}{B};$$

which is maximum + when $v' = 0$, is + till $v' = \frac{B}{4}$ when it vanishes, is − till $v' = \frac{B}{2}$ when it is maximum −, is − till $v' = \frac{3B}{4}$ when it vanishes, and is + till $v' = B$ when it is maximum +.

Therefore the formula of Article 32 will give the following as shewing the nature of the principal alterations in the values of X;

when $v' = 0$, $X = \phi(v)$;

when $v' = \frac{B}{4}$, $X = \phi(v)$ + two positive terms; one depending on D^2, the other on D^3.

when $v' = \frac{B}{2}$, $X = \phi(v)$;

when $v' = \frac{3B}{4}$, $X = \phi(v)$ + a positive term depending on D^2 − a very small positive term depending on D^3.

when $v' = B$, $X = \phi(v)$.

The last very small term of the formula, which vanishes at these five critical points, has between them values which are successively + − − +.

Remarking that the changes in the value of v and v' have the same sign as those of t, so that the smallest value of v' corresponds to the beginning of the displacement of a particle, it will be seen that, in the state of the far-advanced wave,

The first part of the forward displacement is more rapid than in the primitive wave.

The latter part also is more forward, or the retreat is slower, than in the primitive wave.

These peculiarities increase with increase of D.

In these points, the motion of a wave of air is closely analogous to that of a wave of water when its vertical movement is large and it runs for a considerable distance over a shallow bottom. See the *Encyclopædia Metropolitana*, Article, *Tides and Waves*, Section IV., Subsection 3.

34. *Conjectured Change of Character of Wave when it has travelled very far.*

It is difficult to say what will be the form of the wave when $at + x$ is very large. It would be necessary to carry on the steps of the successive substitution to an indefinite extent, or rather, to find a function which would represent the infinite series thus produced. It appears not improbable that at length the continuity of the atmospheric particles may be destroyed, and that something may take place analogous to the bore of a tidal river or the surf of a sea, in which the form and properties of a wave are ultimately lost. (This idea is also suggested by Mr Earnshaw.)

Section IV.

Investigation of the motion of a Wave of Air through the atmosphere considered as of two or three dimensions.

34*. *Outline of the method to be employed; and cautions requiring attention in regard to the order of terms to be rejected.*

The method of forming the equations of motion will be precisely the same, in principle, as that in the instance of air in a tube, Article 21. A symbolical displacement of particles, the most general which the circumstances permit, will be assumed; the symbolical density, and elastic force, and differences of elastic force in different directions, will be found; and these will be compared with the symbolical expressions for changes of velocity which they produce in different directions.

But care is peculiarly necessary, in consequence of the obscurity of the process treating of small terms of a higher order than those which we wish to preserve. In Article 21, we could perceive exactly the form and value of the terms which we rejected: here we can only draw inferences from general reasoning. These inferences, however, will enable us to judge with certainty whether a small quantity before us is of the first order or of a higher order. And it will be seen that quantities of an order which we must retain in one part of the process may be rejected in another part. For instance: the elasticity which determines the motion of a small volume of air is not the absolute elasticity in that volume, but the difference between the elasticity in front and that in rear; and this difference is a small quantity of a higher order than the principal term; but the motion depends *entirely* on it, and it must be carefully retained. But the mass of matter in that volume, to be moved by the differential elasticity, is not the difference between two masses, but is the entire mass in the volume; the difference of the densities in front and in rear is unimportant, and may be wholly rejected. Thus it will be seen that continued attention is necessary for judging on the import of the terms which it is proposed to reject, and on their value as compared with those which it is proposed to retain.

35. *Investigation of the Elastic Force at any point of the disturbed Air.*

Let x, y, z, be the ordinates of a particle in a tranquil state; $x+X$, $y+Y$, $z+Z$, the ordinates of the same particle in its disturbed state at the time t. Conceive seven neighbouring particles forming with the first, in the quiescent state, a rectangular parallelopiped; two bounding planes being defined by the ordinates x and $x+h$, two by the ordinates y and $y+k$, and two by the ordinates z and $z+l$. Confining our attention for a moment to the four particles in a plane parallel to xy, with ordinate z; their original ordinates, and their disturbed ordinates at the time t, parallel to that plane, will be as follows:

Ordinates in the quiescent state,

1st point,	x,	y;
2nd point,	x,	$y+k$;
3rd point,	$x+h$,	y;
4th point,	$x+h$,	$y+k$;

Ordinates in the disturbed state,

$$x+X, \qquad y+Y;$$

$$x+X+\frac{dX}{dy}k, \qquad y+k+Y+\frac{dY}{dy}k;$$

$$x+h+X+\frac{dX}{dx}h, \qquad y+Y+\frac{dY}{dx}h;$$

$$x+h+X+\frac{dX}{dx}h+\frac{dX}{dy}k, \qquad y+k+Y+\frac{dY}{dx}h+\frac{dY}{dy}k.$$

And if, for the disturbed state, we subtract the ordinates of the 1st point from those of 2nd, 3rd, 4th, we have (see Figure 9),

Ordinates relative to the 1st point,

$$\text{2nd point,}\quad \frac{dX}{dy}k, \qquad k+\frac{dY}{dy}k;$$

$$\text{3rd point,}\quad h+\frac{dX}{dx}h, \qquad \frac{dY}{dx}h;$$

$$\text{4th point,}\quad h+\frac{dX}{dx}h+\frac{dX}{dy}k, \quad k+\frac{dY}{dx}h+\frac{dY}{dy}k.$$

It appears from these that the projections on the plane xy, of the four points which were at the angles of a parallelogram, are now at the angles of a lozenge. The ordinates of 1st and 2nd points in the direction of x are not now equal, and those of 1st and 3rd in the direction of y are not now equal. Let the new distance from 1st point to 2nd be p, making the angle ϕ with y; and that from 1st to 3rd be q, making the angle χ with x. Then

$$p.\sin\phi=\frac{dX}{dy}k, \quad p.\cos\phi=k+\frac{dY}{dy}k,$$

$$q.\sin\chi=\frac{dY}{dx}h, \quad q.\cos\chi=h+\frac{dX}{dx}h.$$

The area of the lozenge

$$=pq.\sin(90^\circ-\phi-\chi)$$

$$=pq.\cos\phi.\cos\chi\,(1-\tan\phi.\tan\chi)$$

$$=hk\left(1+\frac{dX}{dx}\right)\left(1+\frac{dY}{dy}\right)(1-\tan\phi.\tan\chi).$$

But tan ϕ is a small quantity containing the multiplier $\frac{dX}{dy}$, and tan χ contains $\frac{dY}{dx}$, and their product is therefore a small quantity of the second order. In our formula we have not included other terms of the second order, and we must therefore reject this product. Hence the area of the projected lozenge, to the first order of terms depending on X and Y, is

$$hk\left(1+\frac{dX}{dx}\right)\left(1+\frac{dY}{dy}\right).$$

Now consider the four points whose ordinates were $z+l$. Upon treating these in the same way, it will be found that at the time t the value of the ordinate of each, parallel to z, is greater than the value of the similar ordinate of the corresponding point among those whose ordinate was z, by $l\left(1+\frac{dZ}{dz}\right)$. To find the solid content of the rhomb, put r for that edge of the rhomb which is nearly parallel to z, s for the small angle which it makes with z, $90°—t$ for its inclination to the lower surface of the rhomb, u for the area of that lower surface, and v for its inclination to the plane xy. The solid content of the rhomb is rigorously $= r \,.\, u \,.\, \cos t$.

Now $r \,.\, \cos s =$ projection of r upon $z = l\left(1+\frac{dZ}{dz}\right)$;

$u \,.\, \cos v =$ projection of u upon $xy = hk\left(1+\frac{dX}{dx}\right)\left(1+\frac{dY}{dy}\right)$.

Putting equivalents for the cosines, we find

$$r.u.\cos t = \frac{hkl\left(1+\frac{dX}{dx}\right)\left(1+\frac{dY}{dy}\right)\left(1+\frac{dZ}{dz}\right)\left(1-2\sin^2\frac{t}{2}\right)}{\left(1-2\sin^2\frac{s}{2}\right)\left(1-2\sin^2\frac{v}{2}\right)}.$$

Now s, t, and v, would have had no existence if the air had not been disturbed, and are produced by the disturbance. They are therefore small quantities of the same order as the disturbance. Consequently $\sin^2\frac{s}{2}$, $\sin^2\frac{t}{2}$, and $\sin^2\frac{v}{2}$, are of the second order of small quantities, and, where they are associated with 1, are to be rejected. But $\frac{dX}{dx}$, $\frac{dY}{dy}$, $\frac{dZ}{dz}$, are of the first order, and, where they are associated with 1, are to be retained. And, finally, we obtain for the solid content of the rhomb,

$$hkl\left(1+\frac{dX}{dx}\right)\left(1+\frac{dY}{dy}\right)\left(1+\frac{dZ}{dz}\right);$$

or, omitting products to the second and higher orders of $\frac{dX}{dx}$, $\frac{dY}{dy}$, $\frac{dZ}{dz}$,

$$hkl\left(1+\frac{dX}{dx}+\frac{dY}{dy}+\frac{dZ}{dz}\right).$$

But the quantity of air at density D, which did occupy the parallelopiped hkl, does now with density Δ occupy the rhomb

$$hkl\left(1+\frac{dX}{dx}+\frac{dY}{dy}+\frac{dZ}{dz}\right).$$

Hence $$\Delta = \frac{D \, . \, hkl}{hkl\left(1 + \frac{dX}{dx} + \frac{dY}{dy} + \frac{dZ}{dz}\right)},$$

which, omitting powers and products of the second and higher orders, gives

$$\Delta = D\left(1 - \frac{dX}{dx} - \frac{dY}{dy} - \frac{dZ}{dz}\right).$$

And, as in Article 12,

$$\Pi = HD\left(1 - \frac{dX}{dx} - \frac{dY}{dy} - \frac{dZ}{dz}\right).$$

This is the elastic pressure about the point whose original ordinates were x, y, z.

36. *Formation of the Equations of Motion.*

In treating of the motions of the particles of air with reference to rectangular co-ordinates, it is necessary to express the forces which act upon small masses of air with reference to those co-ordinates; and therefore, as the elastic force of one portion of the air acts upon the adjacent portion of the air only in a direction normal to the separating surface, we must use separating surfaces parallel to the co-ordinate planes. The surfaces of the lozenge of which we have treated in the last Article are not parallel to the co-ordinate planes, and that lozenge therefore will not suit our present purpose. But we can find *original* values of the three ordinates

$(x+m, y+n, z+p)$ of a particle, which particle, *at the time t*, will have the same values of x and y as the particle whose *original* ordinates were x, y, z. It is evident that if this is done, the two particles so found will, *at the time t*, be separated exactly in the direction of z; the line joining them will be the arète of a parallelopiped whose surfaces are then parallel to the co-ordinate planes; and the equations of motion can be correctly applied from knowledge of the elastic forces corresponding to those particles.

Now, for the particle x, y, z, at time t, x is changed into $x+X$, and y is changed into $y+Y$.

And, for the particle $x+m, y+n, z+p$, at time t, $x+m$ is changed into

$$x+m+X+\frac{dX}{dx}m+\frac{dX}{dy}n+\frac{dX}{dz}p;$$

$y+n$ is changed into

$$y+n+Y+\frac{dY}{dx}m+\frac{dY}{dy}n+\frac{dY}{dz}p.$$

Hence we must have

$$x+X=x+X+\left(1+\frac{dX}{dx}\right)m+\frac{dX}{dy}n+\frac{dX}{dz}p;$$

$$y+Y=y+Y+\frac{dY}{dx}m+\left(1+\frac{dY}{dy}\right)n+\frac{dY}{dz}p;$$

whence $m=-\frac{dX}{dz}p$, and $n=-\frac{dY}{dz}p$, nearly.

Hence, to compare the pressures at two points of the air, which at the time t are separated precisely in the direction of z, we ought to take for the original ordinates of the second point,

$$x-\frac{dX}{dz}p, \quad y-\frac{dY}{dz}p, \quad z+p;$$

and for Π' the pressure, at the time t, about the particle which has come from that second point, we ought to take

$$\Pi-\frac{d\Pi}{dx}\cdot\frac{dX}{dz}p-\frac{d\Pi}{dy}\cdot\frac{dY}{dz}p+\frac{d\Pi}{dz}\cdot p.$$

But the factors $\frac{dX}{dz}$ and $\frac{dY}{dz}$ are small quantities, depending on the extent of motion of the particles. And thus, when, to find the pressure which urges the mass of matter forward, we form $\Pi-\Pi'$, we shall obtain three terms, of which two are smaller than the third, and which, in fact, combined with $\frac{d\Pi}{dx}$ and $\frac{d\Pi}{dy}$, produce quantities of the second order of the particles' motion. We may then neglect them in comparison with the larger term; and thus we have,

$$\Pi-\Pi'=-\frac{d\Pi}{dz}p.$$

The excess of pressure $\Pi-\Pi'$ acts on the base of the rhomb in Article 35, whose area projected on the plane xy is there found

$$=hk\left(1+\frac{dX}{dx}\right)\left(1+\frac{dY}{dy}\right),$$

and therefore the whole excess of pressure urging the rhomb in the direction z is

$$(\Pi - \Pi')\, hk \left(1 + \frac{dX}{dx}\right)\left(1 + \frac{dY}{dy}\right),$$

$$\text{or } -\frac{d\Pi}{dz}\, hkp \left(1 + \frac{dX}{dx}\right)\left(1 + \frac{dY}{dy}\right).$$

Now Π was found in Article 35 to be

$$HD\left(1 - \frac{dX}{dx} - \frac{dY}{dy} - \frac{dZ}{dz}\right),$$

therefore

$$-\frac{d\Pi}{dz} = HD \,.\, \frac{d}{dz}\left(\frac{dX}{dx} + \frac{dY}{dy} + \frac{dZ}{dz}\right).$$

And p in this investigation is the same as l in Article 35. Also, the retention of $\frac{dX}{dx}$ and $\frac{dY}{dy}$ in the external factor would retain terms of the next higher order, which in the formation of Π we have rejected; and therefore they must not be kept here. Thus we obtain for excess of pressure in direction z,

$$HDhkl \,.\, \frac{d}{dz}\left(\frac{dX}{dx} + \frac{dY}{dy} + \frac{dZ}{dz}\right).$$

And the mass to be moved is $Dhkl$.

Hence, by the usual laws of mechanics (Article 21, note), all our densities and pressures being estimated by weights (Article 12), and omitting every consideration of temperature,

$$\frac{d^2(z+Z)}{dt^2} = gH \cdot \frac{d}{dz}\left(\frac{dX}{dx} + \frac{dY}{dy} + \frac{dZ}{dz}\right);$$

or, as z does not depend on t, and $gH = a^2$,

$$\frac{d^2Z}{dt^2} = a^2 \frac{d}{dz}\left(\frac{dX}{dx} + \frac{dY}{dy} + \frac{dZ}{dz}\right).$$

Similarly

$$\frac{d^2Y}{dt^2} = a^2 \frac{d}{dy}\left(\frac{dX}{dx} + \frac{dY}{dy} + \frac{dZ}{dz}\right);$$

$$\frac{d^2X}{dt^2} = a^2 \frac{d}{dx}\left(\frac{dX}{dx} + \frac{dY}{dy} + \frac{dZ}{dz}\right).$$

These three equations express the relation of the motion of the air to the forces producing the motion, in all the directions in which motion can be conceived. They are therefore absolutely sufficient; and no other equation can be introduced, except as equivalent to or deduced from these three.

37. *Introduction of the Characteristic Function F.*

The solution of these equations is in many cases facilitated by the use of a very peculiar Characteristic Function, for which we shall always use the letter F. F is a function of x, y, z, and t. We cannot in all cases find a form of F which shall correspond to an assumed form of solution; but, if we assume the principal characters of a form of F, we can in all cases find the differential equations leading to a solution; and by careful choice of the form, we can usually find solutions possessing the characteristics that we desire.

The definition of the form of F is contained in these assumptions:

$$\frac{dF}{dx}=\frac{dX}{dt}; \quad \frac{dF}{dy}=\frac{dY}{dt}; \quad \frac{dF}{dz}=\frac{dZ}{dt}.$$

Differentiating,

$$\frac{d^2F}{dx^2}=\frac{d^2X}{dt\,.\,dx}; \quad \frac{d^2F}{dy^2}=\frac{d^2Y}{dt\,.\,dy}; \quad \frac{d^2F}{dz^2}=\frac{d^2Z}{dt\,.\,dz}.$$

Therefore,

$$\frac{d^2F}{dx^2}+\frac{d^2F}{dy^2}+\frac{d^2F}{dz^2}=\frac{d}{dt}\left(\frac{dX}{dx}+\frac{dY}{dy}+\frac{dZ}{dz}\right).$$

Now (Article 36),

$$\frac{d^2X}{dt^2}=a^2\frac{d}{dx}\left(\frac{dX}{dx}+\frac{dY}{dy}+\frac{dZ}{dz}\right);$$

but since $\frac{dX}{dt}=\frac{dF}{dx}$, this is changed to

$$\frac{d^2F}{dx\,.\,dt}=a^2\frac{d}{dx}\left(\frac{dX}{dx}+\frac{dY}{dy}+\frac{dZ}{dz}\right).$$

Integrating with respect to x,

$$\frac{dF}{dt}=a^2\left(\frac{dX}{dx}+\frac{dY}{dy}+\frac{dZ}{dz}+\chi(t)\right).$$

Differentiating with respect to t,

$$\frac{d^2F}{dt^2} = a^2 \frac{d}{dt}\left(\frac{dX}{dx} + \frac{dY}{dy} + \frac{dZ}{dz}\right) + \chi'(t);$$

or

$$\frac{d^2F}{dt^2} = a^2\left(\frac{d^2F}{dx^2} + \frac{d^2F}{dy^2} + \frac{d^2F}{dz^2}\right) + \chi'(t);$$

which is the form of equation now to be used. The function $\chi'(t)$ can only produce, in the solution, a function of t, which will vanish in forming $\frac{dF}{dx}$, &c.: it may be omitted without loss of generality.

We should have arrived at the same final result if we had proceeded in the last step from $\frac{d^2Y}{dt^2}$ or $\frac{d^2Z}{dt^2}$.

38. *Inferences from the value of F when its form has been found.*

In Article 35, the density of air at any point is

$$D\left(1 - \frac{dX}{dx} - \frac{dY}{dy} - \frac{dZ}{dz}\right)$$

$$= D\left\{1 - \frac{1}{a^2}\cdot\frac{dF}{dt} + \frac{1}{a^2}\chi(t)\right\}.$$

The existence of terms dependent on t only would imply some general and simultaneous alteration of the density of air in every part of the atmosphere. As this is not consistent with our physical assumptions, we must always suppose F to be so taken that $\chi(t)$ is not required. With this notice, we shall abandon that function, and we have

$$\text{Density} = D\left(1 - \frac{1}{a^2} \cdot \frac{dF}{dt}\right).$$

The motions of a particle are found by the first assumptions,

$$\frac{dX}{dt} = \frac{dF}{dx}; \quad \frac{dY}{dt} = \frac{dF}{dy}; \quad \frac{dZ}{dt} = \frac{dF}{dz}.$$

The disturbed places of the particles are found by integrating these; or

$$X = \int_t \frac{dF}{dx}; \quad Y = \int_t \frac{dF}{dy}; \quad Z = \int_t \frac{dF}{dz}.$$

These expressions, as may be expected, go through some changes in special applications.

39. *Application to a plane wave of air.*

The equation to a plane is,

Normal from origin of co-ordinates

$$= x \cdot \cos\alpha + y \cdot \cos\beta + z \cdot \cos\gamma;$$

where α, β, γ are constant angles, subject to the condition $\cos^2\alpha + \cos^2\beta + \cos^2\gamma = 1$. It seems likely therefore that our object will be gained by supposing F to be a function of the normal $x.\cos\alpha + y.\cos\beta + z.\cos\gamma$. Call this normal w. Then

$$\frac{dF}{dx} = \frac{dF}{dw}\cdot\frac{dw}{dx} = \cos\alpha.\frac{dF}{dw};$$

and

$$\frac{d^2F}{dx^2} = \frac{d}{dx}\left(\frac{dF}{dx}\right) = \frac{d}{dw}\left(\frac{dF}{dx}\right)\cdot\frac{dw}{dx}$$

$$= \frac{d}{dw}\left(\cos\alpha.\frac{dF}{dw}\right).\cos\alpha = \cos^2\alpha.\frac{d^2F}{dw^2}.$$

Similarly

$$\frac{d^2F}{dy^2} = \cos^2\beta.\frac{d^2F}{dw^2}; \quad \frac{d^2F}{dz^2} = \cos^2\gamma.\frac{d^2F}{dw^2}.$$

And

$$\frac{d^2F}{dx^2} + \frac{d^2F}{dy^2} + \frac{d^2F}{dz^2} = \frac{d^2F}{dw^2}.$$

Hence the equation of Article 37 becomes

$$\frac{d^2F}{dt^2} = a^2\frac{d^2F}{dw^2};$$

whose solution is

$$F = \phi\,(at - w) + \psi\,(at + w),$$

$$= \phi\,(at - x.\cos\alpha - y.\cos\beta - z.\cos\gamma)$$

$$+ \psi\,(at + x.\cos\alpha + y.\cos\beta + z.\cos\gamma),$$

$$= \phi\,(at - \text{normal}) + \psi\,(at + \text{normal}).$$

Then, by Article 38,

$$\text{Density} = D\left\{1 - \frac{1}{a}\phi'(at - \text{normal}) - \frac{1}{a}\psi'(at + \text{normal})\right\};$$

$$\frac{dX}{dt} = \frac{dF}{dx} = -\cos\alpha\,.\,\phi'(at - \text{normal}) + \cos\alpha\,.\,\psi'(at + \text{normal});$$

$$X = \frac{-\cos\alpha}{a}\phi(at - \text{normal}) + \frac{\cos\alpha}{a}\psi(at + \text{normal});$$

and similarly for Y and Z.

Everything here depends on the value of the normal upon the plane

$$w = x\,.\cos\alpha + y\,.\cos\beta + z\,.\cos\gamma.$$

This shews that the disturbance of every kind is the same through that plane; and the factors $\cos\alpha$, $\cos\beta$, $\cos\gamma$, in the expressions for X, Y, Z, shew that the movements of the particles in that plane are perpendicular to the plane. The first term exhibits a wave moving, so as to increase the normal, with velocity a; and the second exhibits a wave diminishing the normal with the same velocity.

40. *Combination of two plane waves from two sources.*

Suppose that F consists of two terms G and H, of which G depends on

$$w = x\cos\alpha + y\cos\beta + z\cos\gamma,$$

and H depends on

$$W = x\cos\alpha + y\cos\beta - z\cos\gamma.$$

We shall find, as in last Article,

$$\frac{d^2F}{dt^2} = \frac{d^2G}{dt^2} + \frac{d^2H}{dt^2} = a^2\left(\frac{d^2G}{dw^2} + \frac{d^2H}{dW^2}\right),$$

which, in consequence of our assumptions as to the difference of form of the two terms, will require the separate equations,

$$\frac{d^2G}{dt^2} = a^2\frac{d^2G}{dw^2}, \quad \frac{d^2H}{dt^2} = a^2\frac{d^2H}{dW^2}.$$

Adopting the first wave only in each solution, and taking the same form of function,

$$F = G + H = \phi\,(at - x\cos\alpha - y\cos\beta - z\cos\gamma)$$
$$+ \phi\,(at - x\cos\alpha - y\cos\beta + z\cos\gamma).$$

Therefore, by Article 38,

$$\text{Density} = D\left\{1 - \frac{1}{a}\phi'(at - x\cos\alpha - y\cos\beta - z\cos\gamma)\right.$$

$$\left. - \frac{1}{a}\phi'(at - x\cos\alpha - y\cos\beta + z\cos\gamma)\right\};$$

$$\frac{dZ}{dt} = \frac{dF}{dz} = -\cos\gamma \,.\, \phi'(at - x\cos\alpha - y\cos\beta - z\cos\gamma)$$

$$+ \cos\gamma \,.\, \phi'(at - x\cos\alpha - y\cos\beta + z\cos\gamma);$$

$$Z = -\frac{\cos\gamma}{a}\phi(at - x\cos\alpha - y\cos\beta - z\cos\gamma)$$

$$+ \frac{\cos\gamma}{a}\phi(at - x\cos\alpha - y\cos\beta + z\cos\gamma).$$

And, when $z = 0$,

$$\text{Density} = D\left\{1 - \frac{2}{a}\phi'(at - x\cos\alpha - y\cos\beta)\right\},$$

$$\frac{dZ}{dt} = 0,$$

$$Z = 0,$$

$\frac{dY}{dt}$ and $\frac{dX}{dt}$, however, do not vanish.

It appears therefore that, if the plane xy were a material boundary of the air, the motions of the particles of air would not be altered; since it would permit motion parallel to the plane, and there is no motion perpendicular to the plane. But the density of the air in contact is variable; and therefore, if the barrier completely cuts off communication with tranquil air on the other side, it must be a rigid barrier.

41. *Theory of the simple echo.*

The number of solutions which the equations of Article 36 admit is infinite; embracing not only different forms of function, as in Article 23, but also functions corresponding to waves passing in any different directions in space of three dimensions, to converging waves, to diverging waves, &c. And any one or any combination of these functions, as need requires, may be considered as admissible in quality of the 'undetermined functions' to which attention is called in the author's *Partial Differential Equations*.

If now we wish to ascertain the law of motion of a plane wave, we proceed as in Article 39, and we obtain a solution which shews that the wave preserves its plane character, moving with a certain velocity. And this solution is sufficient, as long as we introduce no condition limiting the space occupied by the air, &c.

But suppose that we introduce this condition, "The air is bounded by an immoveable barrier, constituting the plane xy, and therefore requiring that Z always $= 0$ when $z = 0$." Then our simple solution cannot be made to meet the condition; and we must introduce what is at present an 'undetermined function;' and we must determine it so that, when combined with the simple solution, it shall make $Z = 0$ when $z = 0$.

Since Z in the simple solution

$$= -\frac{\cos\gamma}{a}\phi\,(at - x\cos\alpha - y\cos\beta - z\cos\gamma),$$

the external sign being derived, in the investigation of Article 39, from the sign of z within the bracket; and since the value of Z for $z = 0$ becomes

$$-\frac{\cos\gamma}{a}\phi\,(at - x\cos\alpha - y\cos\beta);$$

it is quickly seen that, for the destruction of this term at all times and with every value of x and y, ϕ must be the same in the new term; the factors of t, x, and y, must be the same; the external factor must be the same but with different sign, and therefore the factor of z, within the bracket for the general value of Z, must be the same but with different sign; and therefore we must add exactly the term added in Article 40. That term represents a wave precisely similar to the first wave, in law and extent of motion of particles, in velocity, and in inclination of its normal to x, y, and z; but differing in this respect, that the inclination of its normal to z is on the opposite side. And therefore, defining the direction of the wave's motion by that normal, the motion of the reflected wave and the motion of the incident wave make equal angles, but on opposite sides, with the normal to the immoveable barrier xy. This is in all respects the character of an Echo of Sound.

42. *Various forms permissible in the expression defining a plane wave of air.*

If, for simplicity of symbols, we suppose the plane of the wave to be parallel to the plane xy, and its motion to be in the direction z, it is easily seen that the equation

$$\frac{d^2F}{dt^2} = a^2\left(\frac{d^2F}{dx^2} + \frac{d^2F}{dy^2} + \frac{d^2F}{dz^2}\right)$$

will be satisfied by the assumption

$$F = K \cdot \phi\,(at - z),$$

provided that K be a function not containing t or z, which satisfies the equation

$$\frac{d^2K}{dx^2} + \frac{d^2K}{dy^2} = 0.$$

This troublesome equation (see the author's *Partial Differential Equations*, Articles 41 and 52) scarcely admits of intelligible solution except under specific assumptions. We may make

$$K = C \cdot xy + D,$$

$$\text{or } K = C\,(x^2 - y^2) + D$$

$$\text{or} \quad K = \epsilon^{nx+C} . \cos (ny + C') + D,$$

$$\text{or} \quad K = C . \log (x^2 + y^2) + D,$$

&c.,

and thus we obtain, as solutions expressing a plane wave of air,

$$F = \{C . xy + D\} . \phi (at - z) ;$$

$$F = \{C (x^2 - y^2) + D\} . \phi (at - z) ;$$

$$F = \{\epsilon^{nx+C} . \cos (ny + C') + D\} . \phi (at - z) ;$$

$$F = \{C . \log (x^2 + y^2) + D\} . \phi (at - z) ;$$

&c. :

and any combination of these with similar terms, or with terms depending on $\psi (at + z)$. On performing the operations of Article 38, it will be seen that these forms imply motion of the particles in the three directions x, y, z, though the motion of the wave is only in z.

It does not appear that these forms have any application in nature.

43. *Remarks on the Partial Differential Equations which occur in the investigations that next follow.*

We shall have to treat equations of the form

$$\frac{1}{a^2} . \frac{d^2 W}{dt^2} = \frac{d^2 W}{dr^2} + \frac{m}{r} . \frac{dW}{dr},$$

with the six values for m, 1, 2, 3, 4, 5, 6. It does not appear that equations of this class can be approached by one general method of attack. Some of them yield to the following. Assume, for trial, $W = \Sigma(A_n . r^n)$; A being in all cases a function of v or $at - r$, which satisfies the equation $\frac{1}{a^2} \cdot \frac{d^2A}{dt^2} = \frac{d^2A}{dr^2}$; and the index n diminishing by successive units. On substituting, and remarking that A' or $\frac{dA}{dv} = -\frac{dA}{dr}$, we find

$$\Sigma \{-(2n+m) . A'_n . r^{n-1} \\ + n(m+n-1) . A_n . r^{n-2}\} = 0,$$

(the accent on A_n denoting the derived function of A_n with regard to v);

and, writing down successive terms instead of the symbol Σ,

$$\left\{\begin{array}{l} \{-(2n+m) . A'_n\} . r^{n-1} \\ + \{-(2n+m-2) . A'_{n-1} + n(m+n-1) . A_n\} . r^{n-2} \\ + \{-(2n+m-4) . A'_{n-2} + (n-1)(m+n-2) . A_{n-1}\} . r^{n-3} \\ + \&c. \end{array}\right\} = 0.$$

[If we suppose A a function of u or $at + r$, the equation obtained is the same, excepting that the signs of the derived functions are changed].

Making each line $= 0$, we find $n = -\frac{m}{2}$, and we find each following function in terms of the preceding function. And the series of functions will terminate in two cases. Either if one of the numbers n, $n-1$, $n-2$, &c. becomes 0; that is, if

$$-\frac{m}{2}, \ -\frac{m}{2}-1, \ -\frac{m}{2}-2, \text{ \&c.,}$$

becomes 0; that is, if $m=0, -2, -4$, &c. Or if one of the numbers $m+n-1$, $m+n-2$, &c., becomes 0; that is, if $\frac{m}{2}-1$, or $\frac{m}{2}-2$, or &c., becomes 0; that is, if $m=2$, or $=4$, &c. The only numbers here which meet our wants are those for $m=2$, $m=4$, $m=6$; and we are still left without solutions for $m=1$, $m=3$, $m=5$. The solution for $m=1$ has, however, been found, as we shall mention, in the unsatisfactory form of a definite integral; and the solutions, when $m=3$, $m=5$, &c., may be made to depend on that when $m=1$.

We invite the attention of the student of Partial Differential Equations to these equations.

For $m=1$, $m=3$, $m=5$, we shall obtain infinite series in descending powers of r, which are practically sufficient for waves diverging to great distances.

For waves nearer to the centre, we may assume

$$W = \Sigma\,(B_n \,.\, r^n),$$

increasing the index by successive units. Treating the series in the same way, we find $n=1-m$, and

$$-(2-m)B'_n \quad + 1 \, . \, (2-m)B_{n+1} = 0,$$

$$-(4-m)B'_{n+1} + 2 \, . \, (3-m)B_{n+2} = 0,$$

&c.

The process however produces the same numbers as the last, and fails for the odd values of m.

44. *Symmetrical divergent wave, with motions of all particles parallel to one plane.*

Assume F to be R, a function of t and r only, where $r = \sqrt{(y^2 + z^2)}$. Then we have,

$$\frac{dF}{dx} = 0\,;$$

$$\frac{dF}{dy} = \frac{dR}{dr} \cdot \frac{dr}{dy} = \frac{dR}{dr} \cdot \frac{y}{\sqrt{(y^2+z^2)}}\,;$$

$$\frac{d^2F}{dy^2} = \frac{d^2R}{dr^2} \cdot \frac{dr}{dy} \cdot \frac{y}{\sqrt{(y^2+z^2)}}$$

$$+ \frac{dR}{dr}\left(\frac{1}{\sqrt{(y^2+z^2)}} - \frac{y^2}{(y^2+z^2)}\right)$$

$$= \frac{d^2R}{dr^2} \cdot \frac{y^2}{y^2+z^2} + \frac{dR}{dr} \cdot \frac{z^2}{(y^2+z^2)^{\frac{3}{2}}}.$$

Similarly

$$\frac{d^2F}{dz^2} = \frac{d^2R}{dr^2} \cdot \frac{z^2}{y^2+z^2} + \frac{dR}{dr} \cdot \frac{y^2}{(y^2+z^2)^{\frac{3}{2}}}.$$

$$\text{The sum} = \frac{d^2R}{dr^2} + \frac{1}{r} \cdot \frac{dR}{dr};$$

and the equation of Article 37 becomes

$$\frac{1}{a^2} \cdot \frac{d^2R}{dt^2} = \frac{d^2R}{dr^2} + \frac{1}{r} \cdot \frac{dR}{dr}.$$

The solution of this equation has been exhibited under the form of the following definite integral; where θ is a new or fictitious angle, introduced solely for the purpose of being the subject of integration; where the integration is to be taken from $\theta = 0$ to $\theta = \pi$; and where S denotes the definite integral so taken:

$$R = S_\theta \,.\, \phi\,(at + r\cos\theta)$$
$$+ S_\theta \,.\, \psi\,(at + r\cos\theta) \,.\, \log\,(r\sin^2\theta).$$

The solution may be verified without much trouble, remarking that differentiations with regard to t and r may be performed under the sign of integration with regard to θ.

It is difficult to extract an intelligible result from this expression. If for $r\cos\theta$ we put $r - 2r\sin^2\frac{\theta}{2}$ or $-r + 2r\cos^2\frac{\theta}{2}$, we can expand in terms of u or v and of integrable functions of θ; but the series proceeds by increasing powers of r, and is unfitted for great distances.

If we use the method of Article 43, we must make

$$m=1,\quad n=-\frac{1}{2};$$

and, supposing

$$R=A_{-\frac{1}{2}}.r^{-\frac{1}{2}}+A_{-\frac{3}{2}}.r^{-\frac{3}{2}}+\&c.,$$

where each of the quantities A is a function of v or $at-r$, implying a diverging wave, the equation of Article 43 becomes

$$\left.\begin{matrix} \left(+2A'_{-\frac{3}{2}}+\frac{1}{2}\cdot\frac{1}{2}A_{-\frac{1}{2}}\right)r^{-\frac{5}{2}} \\ +\left(+4A'_{-\frac{5}{2}}+\frac{3}{2}\cdot\frac{3}{2}A_{-\frac{3}{2}}\right)r^{-\frac{7}{2}} \\ +\&c. \end{matrix}\right\}=0;$$

from which each succeeding A is given by an integral of the preceding A. If we give to A the special form $e.\sin(bv+c)$, these successive integrals are easily expressed; and the series converges with increasing values of r. The same is true if we take for A a function of u, such as $e.\sin(bu+c)$, which implies a converging wave; here, however, as $\frac{dA}{dr}=A'$, the signs of A' must be changed.

45. *Motion of the particles in this problem.*

By Article 38,

$$\frac{dY}{dt} = \frac{dF}{dy}; \quad \frac{dZ}{dt} = \frac{dF}{dz}.$$

The velocity in the direction of radius

$$= \frac{y}{r} \cdot \frac{dY}{dt} + \frac{z}{r} \cdot \frac{dZ}{dt} = \frac{y}{r} \cdot \frac{dF}{dy} + \frac{z}{r} \cdot \frac{dF}{dz};$$

which, substituting the values in the beginning of last Article, is $= \frac{dR}{dr}$. The velocity perpendicular to radius, measured from axis of y towards axis of z,

$$= \frac{y}{r} \cdot \frac{dZ}{dt} - \frac{z}{r} \cdot \frac{dY}{dt} = \frac{y}{r} \cdot \frac{dF}{dz} - \frac{z}{r} \cdot \frac{dF}{dy} = 0.$$

This problem has no remarkable physical application.

46. *Divergent wave, with oscillation of the center of each divergent wave in the direction of z; all motions being parallel to the plane of yz.*

Assume F to be $= R \cdot z$; R being a function of t and r, or $\sqrt{(y^2 + z^2)}$. Then

$$\frac{dF}{dy} = z \cdot \frac{dR}{dy} = z \frac{dR}{dr} \cdot \frac{dr}{dy}$$

$$= z \frac{dR}{dr} \cdot \frac{y}{\sqrt{(y^2 + z^2)}};$$

$$\frac{d^2F}{dy^2} = z\,.\,\frac{d^2R}{dr^2}\,.\,\frac{dr}{dy}\,.\,\frac{y}{\sqrt{(y^2+z^2)}}$$

$$+ z\,.\,\frac{dR}{dr}\left\{\frac{1}{\sqrt{(y^2+z^2)}} - \frac{y^2}{(y^2+z^2)^{\frac{3}{2}}}\right\}$$

$$= z\,.\,\frac{d^2R}{dr^2}\,.\,\frac{y^2}{r^2} + \frac{dR}{dr}\,.\,\frac{z^2}{r^3}.$$

$$\frac{dF}{dz} = R + z\frac{dR}{dz}$$

$$= R + z\,\frac{dR}{dr}\,.\,\frac{z}{\sqrt{(y^2+z^2)}}$$

$$= R + \frac{dR}{dr}\,.\,\frac{z^2}{\sqrt{(y^2+z^2)}};$$

$$\frac{d^2F}{dz^2} = \frac{dR}{dr}\,.\,\frac{z}{\sqrt{(y^2+z^2)}}$$

$$+ \frac{d^2R}{dr^2}\,.\,\frac{z}{\sqrt{(y^2+z^2)}}\,.\,\frac{z^2}{\sqrt{(y^2+z^2)}}$$

$$+ \frac{dR}{dr}\left\{\frac{2z}{\sqrt{(y^2+z^2)}} - \frac{z^3}{(y^2+z^2)^{\frac{3}{2}}}\right\}$$

$$= \frac{dR}{dr}\left(\frac{3z}{r} - \frac{z^3}{r^3}\right) + \frac{d^2R}{dr^2}\,.\,\frac{z^3}{r^2}.$$

$$\frac{d^2F}{dy^2} + \frac{d^2F}{dz^2} = \frac{d^2R}{dr^2}\,z + \frac{dR}{dr}\,.\,\frac{3z}{r}.$$

Also $\dfrac{d^2F}{dt^2} = z\,\dfrac{d^2R}{dt^2}$.

Hence, remarking that $\frac{dX}{dx}$ is 0, and therefore $\frac{d^2F}{dx^2}$ is 0, the equation of Article 37, or

$$\frac{d^2F}{dt^2}=a^2\left(\frac{d^2F}{dy^2}+\frac{d^2F}{dz^2}\right),$$

becomes

$$z.\frac{d^2R}{dt^2}=a^2.z\left(\frac{d^2R}{dr^2}+\frac{3}{r}.\frac{dR}{dr}\right);$$

or

$$\frac{1}{a^2}.\frac{d^2R}{dt^2}=\frac{d^2R}{dr^2}+\frac{3}{r}.\frac{dR}{dr}.$$

To solve this equation, we shall refer to the process of Article 43. There we must make $m=3$, which gives $n=-\frac{3}{2}$; and, supposing the wave to diverge, or B to be a function of v or $at-r$, the equation becomes

$$\left.\begin{array}{l}\left\{2B'_{-\frac{5}{2}}-\frac{3}{2}.\frac{1}{2}B_{-\frac{3}{2}}\right\}r^{-\frac{7}{2}}\\ +\left\{4B'_{-\frac{7}{2}}-\frac{5}{2}.\frac{-1}{2}B_{-\frac{5}{2}}\right\}r^{-\frac{9}{2}}\\ +\&c.\end{array}\right\}=0;$$

where each succeeding coefficient is deduced by an integral of the preceding coefficient. Then R has the form

$$B_{-\frac{3}{2}}.r^{-\frac{3}{2}}+B_{-\frac{5}{2}}.r^{-\frac{5}{2}}+\&c.;$$

and F has the form

$$B_{-\frac{3}{2}}.r^{-\frac{3}{2}}.z+B_{-\frac{5}{2}}.r^{-\frac{5}{2}}.z+\&c.$$

Then the velocity in y, or $\frac{dY}{dt}$, or $\frac{dF}{dy}$, (Article 37), or $\frac{yz}{r}\cdot\frac{dR}{dr}$ (above), can be found; observing that

$$\frac{dB}{dr}=\frac{dB}{dv}\cdot\frac{dv}{dr}=-\frac{dB}{dv}=-B'.$$

And the velocity in z is

$$R+\frac{z^2}{r}\cdot\frac{dR}{dr}.$$

Hence the velocity in the direction of radius

$$=\frac{y}{r}\cdot\frac{yz}{r}\cdot\frac{dR}{dr}+\frac{z}{r}\left(R+\frac{z^2}{r}\cdot\frac{dR}{dr}\right)=\frac{zR}{r}+z\frac{dR}{dr};$$

and the displacement in the direction of radius

$$=\frac{z}{r}\int_t R+z\int_t\frac{dR}{dr}.$$

And the velocity perpendicular to the radius

$$=\frac{y}{r}\left(R+\frac{z^2}{r}\cdot\frac{dR}{dr}\right)-\frac{z}{r}\cdot\frac{yz}{r}\cdot\frac{dR}{dr}=\frac{yR}{r}.$$

47. *Interpretation of the expression for radial displacement in this problem; the particles, originally in a circle whose center is the center of divergence, will always be found in a circle of the same diameter whose center oscillates.*

Suppose that we measure from the center of the wave (or origin of co-ordinates), in the direction z, a distance Q, a function of t and r, which will therefore be the same at any given time for all particles in the circle whose original radius was r, but will vary with the time; and let the distance of that point from the quiescent place of a particle under consideration be called r'. Then $r'^2 = y^2 + (z - Q)^2 = y^2 + z^2 - 2zQ$ nearly $= r^2 - 2zQ$, whence

$$r' = r - \frac{zQ}{r}.$$

The displacement of the particle in the direction of r is

$$\frac{z}{r}\int_t R + z\int_t \frac{dR}{dr}.$$

Hence the disturbed value of r' is

$$r + z\left(\frac{1}{r}\int_t R + \int_t \frac{dR}{dr} - \frac{Q}{r}\right).$$

If now we make

$$Q = \int_t R + r\int_t \frac{dR}{dr},$$

we find

Disturbed value of $r.$ = Undisturbed value of $r.$

That is to say, all the particles, which in the quiescent state were originally in a circle whose center was at the origin, will at the time t be found in a circle of the same diameter whose center has moved through the space

$$\int_t R + r\int_t \frac{dR}{dr},$$

in the direction of z. Hence the wave will be of the character of "divergent wave with oscillation of the center of divergence in the direction of z, the amount of oscillation being different for waves of different diameters."

We leave to the student the investigation of the motion of each particle in the circumference of the circle to which it belongs.

From this it will easily be understood that, if motion be begun in the form of a circle with oscillating center, it will be propagated in the form of a circle with oscillating center; because the general formula expressing disturbance must be such as will represent the special disturbance at the place of beginning of the disturbance, and only the formula that we have found will represent that special disturbance.

48. *Application of this theory to the vibration of air produced by the vibration of musical strings.*

We shall hereafter see that the vibrations of a musical string may always be represented by the combination of vibrations of equal periods, one in one plane passing through the string, and the other in another plane passing through the string, at right angles to the former; and that in the simplest case they will be a vibration in one plane. Confining our attention to the simplest case, and supposing x to be in the direction of the string's length, and z in the direction of vibration, and taking a plane yz at right angles to the wire, it is seen that, for determining the vibrations of air in that plane, we have precisely the case contemplated in the last sentence of last Article, namely, a motion begun in the form of a circle with oscillating center: and the theorems of Articles 46 and 47 apply to it. The principal practical results are: that in the plane xy, or in the plane normal to the plane of vibration, the vibrations of air to and from the wire, on which the audible sound mainly depends, and which here have z for factor, are small; and that, in all directions, the magnitude of vibrations depends principally on R, whose most important term has for factor $r^{-\frac{1}{2}}$, which diminishes rapidly as the distance increases.

49. *Application of the theory to the vibrations of air produced by the vibrations of a tuning-fork.*

The tuning-fork is a small instrument in the form represented in Figure 10, constructed of highly elastic metal. In use, one of its branches is struck, and the whole form is put into a state of vibration; but, by holding the single stalk in the hand, all vibrations of that stalk are speedily deadened, and then (as will be perceived on mechanical principles) the remaining vibrations of the two branches of the fork must be at every moment in opposite directions. As the direction of vibration of each prong of the fork is definite, the theory of Articles 46 and 47 applies to the vibrations of air which it produces; but the combination of the vibrations from the two prongs requires a special investigation. We shall suppose that the section of each stalk is a circle, and that one stalk does not interrupt the air-vibrations produced by the other; neither of which suppositions is strictly true.

Take a plane at right angles to the two prongs for the plane of yz, z passing through both prongs; let the interval between the two prongs be $2c$; take the middle point of that interval for the origin of co-ordinates; put ζ for the ordinate of a particle measured from that point in the direction of z, and ρ for the radial distance from that point; then the z (in our past investigations) of one prong will be $\zeta - c$, and the z of the other prong will be $\zeta + c$. Also the r^2 of one prong will be $(\zeta - c)^2 + y^2 = \zeta^2 + y^2 - 2\zeta c$ nearly, $= \rho^2 - 2\zeta c$, and $r = \rho - \frac{\zeta}{\rho} c$ nearly; for the other prong,

$$r = \rho + \frac{\zeta}{\rho} c \text{ nearly.}$$

The velocity in y therefore produced by one prong, or $\frac{yz}{r} \cdot \frac{dR}{dr}$ (Article 46), is

$$\rho y \frac{\zeta - c}{\rho^2 - \zeta c}\left(\frac{dR}{d\rho} - \frac{d^2R}{d\rho^2} \cdot \frac{\zeta c}{\rho}\right),$$

supposing ρ substituted for r in the function R. Combining the terms, this becomes

$$\frac{y}{\rho^3}\left(\rho^2 \zeta \frac{dR}{d\rho} - y^2 \frac{dR}{d\rho} c - \rho \zeta^2 \frac{d^2R}{d\rho^2} c\right).$$

The velocity in y produced by the other prong would be expressed by the same formula with signs of c changed, if the two prongs were in the same state of vibration. But the two prongs are always in opposite states of vibration. Hence we must again change sign for the entire expression for the movement produced by the other prong. Thus we have

$$\frac{y}{\rho^3}\left(-\rho^2 \zeta \frac{dR}{d\rho} - y^2 \frac{dR}{d\rho} c - \rho \zeta^2 \frac{d^2R}{d\rho^2} c\right);$$

and the entire velocity in the direction of y is,

$$-\frac{2y^3}{\rho^3} \cdot \frac{dR}{d\rho} c - \frac{2y\zeta^2}{\rho^2} \cdot \frac{d^2R}{d\rho^2} c.$$

The velocity in z produced by one prong, or

$$R + \frac{z^2}{r} \cdot \frac{dR}{dr},$$

is $$R - \frac{dR}{d\rho} \cdot \frac{\zeta}{\rho} c + \rho \cdot \frac{\zeta^2 - 2\zeta c}{\rho^2 - \zeta c} \left(\frac{dR}{d\rho} - \frac{d^2R}{d\rho^2} \cdot \frac{\zeta c}{\rho} \right),$$

or $$\frac{\zeta}{\rho^3} \left(\frac{\rho^3}{\zeta} R + \rho^2 \zeta \frac{dR}{d\rho} - 3\rho^2 \frac{dR}{d\rho} c + \zeta^2 \frac{dR}{d\rho} c - \rho\zeta^2 \cdot \frac{d^2R}{d\rho^2} c \right);$$

whence, as before, the entire velocity in z is

$$-\frac{6\zeta}{\rho} \cdot \frac{dR}{d\rho} \cdot c + \frac{2\zeta^3}{\rho^3} \cdot \frac{dR}{d\rho} c - \frac{2\zeta^3}{\rho^2} \cdot \frac{d^2R}{d\rho^2} c;$$

and the total velocity in the direction of ρ

$$= \frac{y}{\rho} \times \text{velocity in } y + \frac{\zeta}{\rho} \times \text{velocity in } \zeta$$

$$= \frac{-2c}{\rho^2} \left\{ (\rho^2 + \zeta^2) \frac{dR}{d\rho} + \zeta^2 \rho \frac{d^2R}{d\rho^2} \right\}.$$

Without going into the complete discussion of this formula, it may be stated that in many cases $\frac{d^2R}{d\rho^2}$ has a sign opposite to that of $\frac{dR}{d\rho}$, so that the coefficient of ζ^2 is negative; and with certain values of ζ, that is, in a direction making a certain angle with y, there will be no vibration. This may be verified by putting a tuning-fork into vibration, holding it to the ear, and turning it round its stalk: in four positions the sound sensibly dies away.

The coefficient of vibration of air, which is proportional to r^{-3} (see the expression for R in Article 46), decreases rapidly with increasing distance.

50. *Symmetrical divergent wave, in air of three dimensions; convenience of circnlar functions for expression of wave: modification of the magnitude and law of the displacement of particles, and of the speed of wave, as the distance from the origin increases.*

Assume F to be R, a function of t and r only, where

$$r = \sqrt{(x^2 + y^2 + z^2)}.$$

Proceeding exactly as in Article 44, we find

$$\frac{dF}{dx} = \frac{dR}{dr} \cdot \frac{dr}{dx} = \frac{dR}{dr} \cdot \frac{x}{\sqrt{(x^2 + y^2 + z^2)}};$$

$$\frac{d^2F}{dx^2} = \frac{d^2R}{dr^2} \cdot \frac{dr}{dx} \times \frac{x}{\sqrt{(x^2 + y^2 + z^2)}}$$

$$+ \frac{dR}{dr} \times \left\{\frac{1}{\sqrt{(x^2 + y^2 + z^2)}} - \frac{x^2}{(x^2 + y^2 + z^2)^{\frac{3}{2}}}\right\}$$

$$= \frac{d^2R}{dr^2} \cdot \frac{x^2}{x^2 + y^2 + z^2} + \frac{dR}{dr} \cdot \frac{y^2 + z^2}{(x^2 + y^2 + z^2)^{\frac{3}{2}}}.$$

Similarly,

$$\frac{d^2F}{dy^2} = \frac{d^2R}{dr^2} \cdot \frac{y^2}{x^2 + y^2 + z^2} + \frac{dR}{dr} \cdot \frac{x^2 + z^2}{(x^2 + y^2 + z^2)^{\frac{3}{2}}},$$

$$\frac{d^2F}{dz^2} = \frac{d^2R}{dr^2} \cdot \frac{z^2}{x^2 + y^2 + z^2} + \frac{dR}{dr} \cdot \frac{x^2 + y^2}{(x^2 + y^2 + z^2)^{\frac{3}{2}}}.$$

And the equation of Article 37 becomes

$$\frac{1}{a^2}\cdot\frac{d^2R}{dt^2}=\frac{d^2R}{dr^2}+\frac{2}{r}\cdot\frac{dR}{dr}.$$

This equation can be completely integrated by the process of Article 43. Making $m=2$, $n=-1$, and assuming the solution to be

$$C_{-1}.r^{-1}+C_{-2}.r^{-2}+C_{-3}.r^{-3}+\&c.,$$

we find C_{-2} and all succeeding coefficients $=0$; and the value of R is simply $\frac{C}{r}$, where C is a function of u and v.

As in Article 45, the velocity of a particle in the direction of the radius vector $=\frac{dR}{dr}$; and if the wave be a divergent wave, C is a function of v only, or of $at-r$;

$$\frac{dC}{dr}=\frac{dC}{dv}\cdot\frac{dv}{dr}=-\frac{dC}{dv}=-C';$$

and the velocity of a particle, or $\frac{dR}{dr}$

$$=-\frac{C'}{r}-\frac{C}{r^2}.$$

Integrating this quantity with regard to t, and making

$$\frac{dD}{dv}=C,$$

the displacement of a particle is found to be

$$-\frac{D'}{ar}-\frac{D}{ar^2}.$$

The import of this expression will best be seen by assuming a form for the function D. Suppose

$$D=b\,.\,\sin\frac{2\pi}{\lambda}(at-r),$$

the same form of function which is found convenient in the Undulatory Theory of Light for a series of similar waves in continued succession, and to which (as we shall hereafter shew) all systems of recurring similar waves may be referred. The value of this function, after going through various changes with continued increase in the value of r, returns, when r is increased by λ, to the same value as before; and as the characteristic of a series of similar recurring waves is that at the end of a certain spatial interval, which we call "the length of a wave," the state of disturbance is the same as at the beginning of that interval, it follows that λ is the length of a wave. The same recurrence of value is produced with unaltered r if we increase t by $\frac{\lambda}{a}$; which shews that the wave advances through a space equal to its length, so as to leave a succeeding wave exactly in the same place in which the preceding wave was, in the time $\frac{\lambda}{a}$.

Now, since

$$D = b \,.\, \sin \frac{2\pi}{\lambda}(at - r), \quad \text{or} = b \,.\, \sin \frac{2\pi}{\lambda} v,$$

D' will be

$$\frac{2\pi}{\lambda} b \,.\, \cos \frac{2\pi}{\lambda} v;$$

and the expression for displacement is

$$-\frac{b}{a}\left(\frac{1}{r}\,\frac{2\pi}{\lambda} \cos \frac{2\pi}{\lambda} v + \frac{1}{r^2} \sin \frac{2\pi}{\lambda} v\right),$$

or $$-\frac{b}{ar^2}\left(\frac{2\pi r}{\lambda} \cos \frac{2\pi}{\lambda} v + \sin \frac{2\pi}{\lambda} v\right).$$

If we make $\tan \theta = \frac{2\pi r}{\lambda}$, this expression becomes

$$-\frac{b}{ar^2} \cdot \frac{1}{\cos \theta} \sin \left(\frac{2\pi v}{\lambda} + \theta\right),$$

or $$-\frac{2\pi b}{a\lambda r} \sqrt{\left(1 + \frac{\lambda^2}{4\pi^2 r^2}\right)} \,.\, \sin \left(\frac{2\pi v}{\lambda} + \theta\right).$$

The angle θ increases from 0 (when $r = 0$) to $\frac{\pi}{2}$ (when r is ∞). Thus we find that,

The displacement of the particles is expressed by a modified wave, in which the maximum of backwards-and-forwards disturbance is not the same at all distances from the center of divergence, but varies more rapidly than the reciprocal of distance from that center. For great distances, however, it is proportional to the reciprocal of distance.

The progress of the modified wave is not uniform; for to the quantity $\frac{2\pi v}{\lambda}$ there is attached θ, or to v there is attached $\frac{\lambda\theta}{2\pi}$, a quantity increasing positively but more rapidly at first than at last. Conceive this united with $-r$, and let $r-\frac{\lambda\theta}{2\pi}=r'$. Then the last factor of displacement is

$$\sin\frac{2\pi}{\lambda}(at-r').$$

This shews that at the time t the multiple at in v is connected with r', a quantity smaller than r. To ascertain the spatial interval of waves at a given time τ, we must change r' by such a quantity that $\frac{2\pi}{\lambda}r'$ will be changed by 2π, that is, r' will be changed by λ, or $r-\frac{\lambda\theta}{2\pi}$ will be changed by λ, or r will be changed by

$$\lambda+\frac{\lambda\theta}{2\pi}.$$

Hence the spatial interval of the waves is rendered rather larger by this term; the interval in time being, at any given point, necessarily unaltered (as determined only by the interval of impulses at the origin), and the velocity of the waves is a little increased. But the whole gain in space travelled over by a wave is

$$\frac{\lambda}{2\pi} \times \text{whole gain in } \theta = \frac{\lambda}{2\pi} \times \frac{\pi}{2} = \frac{\lambda}{4}.$$

51. *Divergent wave of three dimensions, with oscillation of the center of divergence in the direction of z.*

Assume F to be Rz, R being a function of t and r only, where

$$r = \sqrt{(x^2 + y^2 + z^2)}.$$

Then, by a process exactly similar to that of Art. 46,

$$\frac{dF}{dx} = z \cdot \frac{dR}{dx} = z \cdot \frac{dR}{dr} \cdot \frac{dr}{dx} = z \cdot \frac{dR}{dr} \cdot \frac{x}{\sqrt{(x^2 + y^2 + z^2)}};$$

$$\frac{d^2F}{dx^2} = z \cdot \frac{d^2R}{dx^2} = \frac{d^2R}{dr^2} \cdot \frac{x^2 z}{x^2 + y^2 + z^2} + \frac{dR}{dr} \cdot \frac{y^2 z + z^3}{(x^2 + y^2 + z^2)^{\frac{3}{2}}},$$

(by the expansion of last Article).

Similarly,

$$\frac{d^2F}{dy^2} = \frac{d^2R}{dr^2} \cdot \frac{y^2 z}{x^2 + y^2 + z^2} + \frac{dR}{dr} \cdot \frac{x^2 z + z^3}{(x^2 + y^2 + z^2)^{\frac{3}{2}}}.$$

Then

$$\frac{dF}{dz} = R + z \,.\, \frac{dR}{dz} = R + \frac{dR}{dr} \,.\, \frac{z^2}{\sqrt{(x^2+y^2+z^2)}} \,;$$

$$\frac{d^2F}{dz^2} = \frac{dR}{dr} \,.\, \frac{z}{\sqrt{(x^2+y^2+z^2)}} + \frac{d^2R}{dr^2} \,.\, \frac{z^3}{x^2+y^2+z^2}$$

$$+ \frac{dR}{dr} \,.\, \frac{2x^2z + 2y^2z + z^3}{(x^2+y^2+z^2)^{\frac{3}{2}}} \,,$$

$$= \frac{d^2R}{dr^2} \,.\, \frac{z^3}{x^2+y^2+z^2} + \frac{dR}{dr} \,.\, \frac{3x^2z + 3y^2z + 2z^3}{(x^2+y^2+z^2)^{\frac{3}{2}}} \,.$$

The sum $\frac{d^2F}{dx^2} + \frac{d^2F}{dy^2} + \frac{d^2F}{dz^2}$ is

$$\frac{d^2R}{dr^2} \,.\, z + \frac{dR}{dr} \,.\, \frac{4z}{r} \,.$$

And

$$\frac{d^2F}{dt^2} = z \,.\, \frac{d^2R}{dt^2} \,.$$

Thus the equation of Article 37 becomes

$$\frac{1}{a^2} \,.\, \frac{d^2R}{dt^2} = \frac{d^2R}{dr^2} + \frac{4}{r} \,.\, \frac{dR}{dr} \,.$$

To solve this equation, we adopt the process of Article 43.

Making $n = -2$, and using the form

$$E_{-2} \,.\, r^{-2} + E_{-3} \,.\, r^{-3} + \&c.,$$

we have

$$2E'_{-3} - 2 \,.\, 1 \,.\, E_2 = 0,$$

$$4E'_{-4} - 3 \,.\, 0 \,.\, E_{-3} = 0 \,;$$

whence E'_{-4} and all following quantities are $=0$, and $E'_{-3}=E_{-4}$. If $E_{-2}=\frac{dG}{dv}$, then $E_{-3}=G$, and

$$R=\frac{dG}{dv}\cdot\frac{1}{r^2}+\frac{G}{r^3},$$

and

$$F=z\left(\frac{dG}{dv}\cdot\frac{1}{r^2}+\frac{G}{r^3}\right).$$

This, it will be remembered, is on the supposition that we confine ourselves to the divergent wave, and consider G as a function of v or $at-r$ only; if, to complete the solution, we desire also to introduce the convergent wave, we must introduce H a function of u or $at+r$ only, and the sign of its differential coefficient must be changed.

The velocity in x, or $\frac{dF}{dx}$ (Article 38), is

$$z\frac{dR}{dx}=\frac{zx}{r}\cdot\frac{dR}{dr}.$$

The velocity in y

$$=\frac{zy}{r}\cdot\frac{dR}{dr}.$$

The velocity in z

$$=R+\frac{z^2}{r}\cdot\frac{dR}{dr}.$$

The velocity in the direction of $r=\frac{x}{r}\times$ velocity in x

$+\frac{y}{r}\times$ velocity in $y+\frac{z}{r}\times$ velocity in $z=\frac{zR}{r}+z\frac{dR}{dr}$.

The displacement in the direction of r

$$=\frac{z}{r}\int_t R+z\int_t \frac{dR}{dr}.$$

From this it follows, exactly as in Article 47, that the particles originally in a spherical surface whose center is the center of divergence will always be in a spherical surface of the same diameter with oscillating center. And the displacement of the center of the spherical surface at the time t is $\int_t R+r\int_t \frac{dR}{dr}$ in the direction of z.

The elastic pressure of the air at any point (Article 35) is $HD\left(1-\frac{dX}{dx}-\frac{dY}{dy}-\frac{dZ}{dz}\right)$, which (Article 38) is the same as

$$HD\left(1-\frac{1}{a^2}\cdot\frac{dF}{dt}\right),$$

or

$$HD\left(1-\frac{z}{a^2}\cdot\frac{dR}{dt}\right).$$

52. *Application of this theory to an oscillating pendulum with spherical bob; first, symbolical integration for the pressure on the whole surface.*

A spherical solid body moving in the direction of z will have for its coating of air one of the spherical surfaces of which we have spoken, and from it the waves will diverge according to the law found in the last Article; the necessary condition being, that the displacement of center of that air-coating must be the same as the displacement of the center of the solid sphere, or that the function R must be so determined that, when applied to the sphere of the same radius as the solid sphere, $\int_t R + r\int_t \frac{dR}{dr}$ must be equal to the displacement of the center of the solid sphere.

Now let us examine the value of the elastic pressure upon the surface of the solid sphere.

First, we will examine the pressure upon any spherical surface whose radius is r, and whose center was originally at the origin of co-ordinates.

Conceive the surface of the sphere divided into annuli by planes parallel to xy; let two of these planes be at distances from the center of the solid sphere z and $z+\delta z$. (This z and δz are not exactly the same as the original z and δz, because the particles of air have a motion in each spherical surface; but the difference depends on the first order of displacements; and its effect on the variable part of the elastic force, which itself is of the first order, will be of the second order, and may be neglected.) If we put ξ for $\sqrt{(x^2+y^2)}$, so that

$$\xi^2+z^2=r^2,$$

which for this investigation is constant, ξ and $\xi+\delta\xi$ will be the radii of the circular intersections by the two planes. The resolved part of the elastic force which retards the motion of the sphere in the direction of z is $=$ elastic force $\times$ area included between circles of diameters ξ and $\xi+\delta\xi$; but as, with positive and increasing z, ξ diminishes, we must say that the resolved part of the elastic force which accelerates the motion of the sphere $=$ elastic force $\times 2\pi\xi\delta\xi = -$ elastic force $\times 2\pi z\delta z$

$$= 2\pi HD\left(-1+\frac{z}{a^2}\cdot\frac{dR}{dt}\right)z\delta z.$$

We arrive at the same expression if we confine our attention to that side of the sphere where z is negative.

Therefore, to find the total pressure acting to accelerate the sphere in the direction of z, we must integrate the quantity

$$2\pi HD\left(-z+\frac{z^2}{a^2}\cdot\frac{dR}{dt}\right)$$

with respect to z, from $z=-r$ to $z=+r$.

The first term produces 0.

The second term produces

$$+\frac{4\pi}{3}HD\cdot\frac{r^3}{a^2}\cdot\frac{dR}{dt};$$

or, since

$$\frac{dR}{dt}=\frac{dR}{dv}\cdot\frac{dv}{dt}=a\cdot\frac{dR}{dv},$$

the second term produces

$$+\frac{4\pi}{3}\cdot\frac{HD}{a}\left(r\frac{d^2G}{dv^2}+\frac{dG}{dv}\right).$$

Now the ordinate Q of the center of the sphere is

$$\int_t \left(R + r\frac{dR}{dr}\right);$$

therefore

$$\frac{dQ}{dt} = R + r\frac{dR}{dr};$$

or, as $R = \frac{dG}{dv} \cdot \frac{1}{r^2} + \frac{G}{r^3}$, where G is a function of v or $(at - r)$ (Article 51),

$$\frac{dQ}{dt} = -\frac{d^2G}{dv^2} \cdot \frac{1}{r} - 2\frac{dG}{dv} \cdot \frac{1}{r^2} - \frac{2G}{r^3}.$$

53. *Pendulum investigation continued; determination of the form of the function, and evaluation of the entire pressure.*

In the case of a vibrating pendulum, whose bob is a sphere of radius ρ, the ordinate Q, of the center of the bob or of the atmospheric sphere whose radius is ρ, will move according to this law,

When ρ is put for r, Q must $= b \cdot \sin ct$.

Therefore the function G must be so taken that $\frac{dQ}{dt}$, as expressed above, will, when ρ is put for r, assume the shape $bc \cdot \cos ct$. There is no hope of doing this except by making G a simple function of sine and cosine. Suppose then we assume for G,

$$K.\text{sine (multiple of } v + \text{constant)}$$
$$+ L.\text{cosine (multiple of } v + \text{constant)}.$$

The multiple of v must be such that, to make our terms comparable with $\cos ct$, the multiple of t under the bracket will be c. But $v = at - r$. Hence the multiple of v will be $\frac{c}{a}$; and we have for G,

$$K.\text{sine}\left\{\frac{c}{a}(at - r) + \text{constant}\right\}$$
$$+ L.\text{cosine}\left\{\frac{c}{a}(at - r) + \text{constant}\right\}.$$

When for r we put ρ, this becomes

$$k.\text{sine}\left\{\frac{c}{a}(at - \rho) + \text{constant}\right\}$$
$$+ l.\text{cosine}\left\{\frac{c}{a}(at - \rho) + \text{constant}\right\};$$

where k and l are the values which K and L receive when for r we put the special value ρ.

But in that case, which is to correspond to the motion of the center of the pendulum-bob, our terms are to depend on ct. Therefore the constant, in each term, must be $+\frac{c\rho}{a}$; and our general expression must be,

$$G = K.\text{sine}\frac{c}{a}(v + \rho) + L.\text{cosine}\frac{c}{a}(v + \rho).$$

Substituting this in the expression for $\frac{dQ}{dt}$, which, when after the differentiation ρ is put for r, is to equal $bc.\cos ct$, and remarking that for the general symbols K and L we are now to put the special values k and l,

$$\left[\begin{array}{l}\left\{\left(\frac{c^2}{a^2\rho}-\frac{2}{\rho^3}\right)k+\frac{2c}{a\rho^2}l\right\}\sin ct\\ +\left\{\left(\frac{c^2}{a^2\rho}-\frac{2}{\rho^3}\right)l-\frac{2c}{a\rho^2}k\right\}\cos ct\end{array}\right]=+bc.\cos ct.$$

Put e and f for the coefficients of k and l in the first line; then we obtain

$$k=\frac{-bcf}{e^2+f^2},\qquad l=\frac{+bce}{e^2+f^2}.$$

Forming with these the general expression for

$$\frac{4\pi}{3}\cdot\frac{HD}{a}\left(r\frac{d^2G}{dv^2}+\frac{dG}{dv}\right),$$

which (Article 52) gives the pressure in direction z upon the surface of the sphere whose radius is r; and putting ρ for r; we have for the whole pressure on the solid sphere in the direction of z,

$$\frac{4\pi}{3}\cdot\frac{HDbc^2}{a^5\rho^3(e^2+f^2)}\{(a\rho^2c^2+2a^3)\sin ct-c^3\rho^3.\cos ct\}.$$

Substituting for e and f their values, the whole pressure on the sphere in the direction of z is

$$\frac{4\pi}{3}\rho^3 D \cdot \frac{H}{a} \cdot \frac{bc^3}{c^4\rho^4 + 4a^4}\{(a\rho^2c^2 + 2a^3)\sin ct - c^3\rho^3 \cos ct\}..$$

And it is to be remarked that $\frac{4\pi}{3}\rho^3 D =$ mass of air displaced by the solid sphere.

Now bc, the maximum velocity of the pendulum, is very small in comparison with a the velocity of sound. Retaining only the principal term,

pressure in the direction of z

$$= \text{mass of displaced air} \times \frac{H}{2a^2} \cdot bc^2 \cdot \sin ct.$$

54. *Pendulum investigation completed; comparison of the atmospheric pressure with the resolved part of gravity; the oscillations of the pendulum are made to occupy a longer time.*

Let W be the weight of the bob. Since $Q = b \cdot \sin ct$,

$$\frac{d^2Q}{dt^2} = -bc^2 \cdot \sin ct,$$

and the resolved pressure arising from gravity which produces the vibration of W is

$$\frac{-W \cdot bc^2 \cdot \sin ct}{g}.$$

Hence the proportion of the pressure in z produced by the elasticity of air, to the pressure in z produced by the resolved part of gravity, is

$$\frac{-\text{weight of displaced air} \times Hg}{2a^2 \times \text{weight of bob}}.$$

But in Article 22, $Hg = a^2$.

Hence the proportion of the pressure in the direction of motion produced by air-elasticity, to the pressure produced by gravity, is

$$-\frac{1}{2}\cdot\frac{\text{weight of displaced air}}{\text{weight of pendulum-bob}}.$$

This new pressure, it is to be remarked, diminishes the effect of gravity. It is not of the nature of friction, which depends only on the velocity; (the very small term multiplying cos ct, which is proportional to the velocity, is of the nature of friction). But the term which we have considered may be represented by saying, either that gravity is diminished, or that the inertia of the pendulum-bob is increased by the addition of a mass of air equal in volume to half the volume of the pendulum-bob. The latter is the more usual form of expressing the result.

The effect of it is, that the pendulum is made thereby to vibrate more slowly.

It is particularly to be remarked that this retarding effect is totally different from that arising from the "floatation" of the pendulum-bob in air. The effect of that floatation is to diminish the acting weight of the bob by the whole weight of displaced air, and to diminish the power of gravity in producing vibration by the fraction

$$\frac{\text{weight of displaced air}}{\text{weight of pendulum-bob}}.$$

Hence the total diminution of the power of gravity in producing vibration is expressed by the fraction

$$\frac{3}{2} \cdot \frac{\text{weight of displaced air}}{\text{weight of pendulum-bob}},$$

or

$$\frac{3}{2} \cdot \frac{\text{specific gravity of air}}{\text{specific gravity of pendulum-bob}}.$$

This proposition is very important in the computation of pendulum experiments.

As with unaltered gravity, so with this altered gravity, there is no tendency to diminish the arc of vibration.

54*. *Divergent wave of three dimensions; each ring of particles, which was originally in a circle parallel to* xy, *changing its form into an ellipse; and the major and minor axes of the ellipse alternating in direction.*

Assume F to be Rxy, where R is a function only of t and r or $\sqrt{(x^2 + y^2 + z^2)}$. Performing the differentiations exactly as in the last investigations, we find

$$\frac{d^2F}{dx^2} = xy \cdot \frac{d^2R}{dr^2} \cdot \frac{x^2}{r^2} - \frac{xy}{r} \cdot \frac{dR}{dr} \cdot \frac{x^2}{r^2} + \frac{3xy}{r} \cdot \frac{dR}{dr},$$

$$\frac{d^2F}{dy^2} = xy \cdot \frac{d^2R}{dr^2} \cdot \frac{y^2}{r^2} - \frac{xy}{r} \cdot \frac{dR}{dr} \cdot \frac{y^2}{r^2} + \frac{3xy}{r} \cdot \frac{dR}{dr},$$

$$\frac{d^2F}{dz^2} = xy \cdot \frac{d^2R}{dr^2} \cdot \frac{z^2}{r^2} - \frac{xy}{r} \cdot \frac{dR}{dr} \cdot \frac{z^2}{r^2} + \frac{xy}{r} \cdot \frac{dR}{dr};$$

$$\frac{d^2F}{dx^2} + \frac{d^2F}{dy^2} + \frac{d^2F}{dz^2} = xy \left\{ \frac{d^2R}{dr^2} + \frac{6}{r} \cdot \frac{dR}{dr} \right\}.$$

Also $\dfrac{d^2F}{dt^2} = xy \cdot \dfrac{d^2R}{dt^2}$;

Hence the equation of Article 37 becomes

$$\frac{1}{a^2}\cdot\frac{d^2R}{dt^2}=\frac{d^2R}{dr^2}+\frac{6}{r}\cdot\frac{dR}{dr}.$$

By Article 43, a solution of this equation can be found in the form

$$R=K_{-3}\,.\,r^{-3}+K_{-4}\,.\,r^{-4}+K_{-5}\,.\,r^{-5},$$

where K_{-3}, K_{-4}, K_{-5}, are functions of v or of $at-r$. Remarking that

$$\frac{d\,.\,K_{-3}}{dt}=\frac{d\,.\,K_{-3}}{dv}\cdot\frac{dv}{dt}=a\,.\,\frac{d\,.\,K_{-3}}{dv},$$

and that

$$\frac{d\,.\,K_{-3}}{dr}=\frac{d\,.\,K_{-3}}{dv}\cdot\frac{dv}{dr}=-\frac{d\,.\,K_{-3}}{dv},$$

and so for each of the other functions; our equation becomes on substitution,

$$0=\left\{-6\,.\,K_{-3}+2\frac{d\,.\,K_{-4}}{dv}\right\}r^{-5}+\left\{-4K_{-4}+4\frac{d\,.\,K_{-5}}{dv}\right\}r^{-6}.$$

Making each term $=0$, and supposing $K_{-5}=H$, a function of v,

$$K_{-4}=\frac{dH}{dv},\quad K_{-3}=\frac{1}{3}\cdot\frac{d^2H}{dv^2};$$

and $$F=xyR=xy\left\{\frac{1}{3}\cdot\frac{d^2H}{dv^2}r^{-3}+\frac{dH}{dv}r^{-4}+Hr^{-5}\right\}.$$

54**. *Motions of the particles in this problem: similarity of the motions to those of the surface of a bell.*

By Article 37,

$$\frac{dX}{dt}=\frac{dF}{dx}=\frac{d}{dx}(xyR)=yR+xy\frac{dR}{dr}\cdot\frac{x}{r}.$$

Similarly $\frac{dY}{dt} \qquad = xR+xy\frac{dR}{dr}\cdot\frac{y}{r}.$

The velocity parallel to xy in the direction of radius

$$=\frac{x}{r}\cdot\frac{dX}{dt}+\frac{y}{r}\cdot\frac{dY}{dt}=2xy.\frac{R}{r}+xy.\frac{dR}{dr}=xy\left(\frac{2R}{r}+\frac{dR}{dr}\right).$$

This expression vanishes when $x=0$, and also when $y=0$. That is, supposing the plane xy to be horizontal, in two vertical planes intersecting at the origin of co-ordinates there is no motion whatever to or from the center. But in the first quadrant of azimuth, with positive x and positive y, the sign of velocity in radius is the same as the sign of the bracket; in the second quadrant, with negative x and positive y, the sign is opposite to that of the bracket; in the third quadrant, where x and y are both negative, the sign is the same as that of the bracket; and in the fourth quadrant, where x is positive and y negative, the sign is opposite to that of the bracket. This shews that the form of a series of particles, which in the quiescent state was circular, becomes elliptic, the axes of the ellipse being inclined 45° to the axes of x and y. And if, as in Article 50, we suppose H to be a periodical function of v, alternately + and − in successive small portions of time; then the quantity in the bracket will be alternately + and − in successive small portions of time; in these successive small portions of time, the radial velocity in the first quadrant will change from + to − while that in the second quadrant changes from − to + (or *vice versâ*), and so for the other quadrants. Therefore the form of the series of particles, originally circular, will be changing backwards and forwards, from an ellipse with major axis inclined 45° to x, to an ellipse with major axis inclined 135° to x.

This motion is exactly that of the particles of a bell, hung vertically, and struck by a hammer (exterior or interior) on one side. And, the air being set in motion in this form, it will (in consequence of the applicability of the formula to all values of r) propagate its motion, as regards radial movement, in the same form. The amplitude of the movement, or the coefficient of the periodical term, has for its principal factor

$$xy \cdot \frac{R}{r}, \text{ or } xy \cdot \frac{d^2H}{dv^2} \cdot r^{-4};$$

which, for a given azimuth, is inversely as the square of the distance.

The velocity transverse to the radius is

$$\frac{x}{r} \cdot \frac{dY}{dt} - \frac{y}{r} \cdot \frac{dX}{dt} = (x^2 - y^2)\,\frac{R}{r}.$$

This is maximum where the radial velocity vanishes and vanishes where the radial velocity is maximum.

The velocity in the direction of z, which is $\frac{dF}{dz}$, or $\frac{d}{dz}(xyR)$, is simply $xy\,\frac{dR}{dr} \cdot \frac{z}{r}$, or $\frac{xyz}{r} \cdot \frac{dR}{dr}$.

The theory applies to the external movements of air produced by a spherical bell struck on one side, or by a hemispherical bell whose mouth is close to the ground.

54***. *Divergent wave of two dimensions, of the same character as the last.*

Treating this in the same manner as in Articles 44 and 46, with the assumption $F = Rxy$, where R is a function of t and of $\sqrt{(x^2 + y^2)}$, the equation for R is found to be

$$\frac{1}{r^2} \cdot \frac{d^2R}{dt^2} = \frac{d^2R}{dr^2} + \frac{5}{r} \cdot \frac{dR}{dr}.$$

It does not appear necessary to follow the consequences of this equation in detail.

55. *Equations for a horizontal plane wave passing upwards through the atmosphere, the effect of gravity being taken into account.*

In all the investigations to this point we have treated the air as an elastic fluid, confined so as to be prevented from disseminating itself into infinite space, but not subject to the action of gravity. We shall now consider, in a simple case, the modifications which are introduced by the introduction of gravity.

First, we will find the relation between the pressure Π, the density Δ, and the height x, while the air is in its quiescent state. In a vertical column whose section is 1, the mass between x and $x+\delta x$ is $\Delta\times\delta x$; and this is supported by the excess of the pressure Π below it over the pressure $\Pi+\delta\Pi$ above it. Therefore

$$\Delta\times\delta x=-\delta\Pi\,;$$

whence, taking the limiting value of the fraction,

$$\frac{d\Pi}{dx}=-\Delta.$$

By Article 13, $\Delta=\frac{\Pi}{H}$; therefore

$$\frac{1}{\Pi}\cdot\frac{d\Pi}{dx}=\frac{-1}{H};$$

whence

$$\log\Pi=\frac{-x}{H}+C;$$

and, putting P for the pressure at the ground, $\log P=C$; therefore

$$\log\frac{\Pi}{P}=-\frac{x}{H},\quad\text{or } \Pi=P\,.\,\epsilon^{-\frac{x}{H}};$$

and

$$\Delta=\frac{P}{H}\,.\,\epsilon^{-\frac{x}{H}}.$$

Now suppose the particle at elevation x to be raised to the elevation $x+X$, and that at elevation $x+h$ to be raised to

$$x+h+X+\frac{dX}{dx}h.$$

As in preceding instances, the new density Δ' will

$$=\Delta\left(1-\frac{dX}{dx}\right).$$

The pressure about the particle whose original place was x being Π', that about the particle whose original place was $x+k$ will be

$$\Pi' + \frac{d\Pi'}{dx} k;$$

the excess of upwards pressure is $-\dfrac{d\Pi'}{dx} k$.

The weight of the matter in the same space, pressing it downwards (by the action of gravity), is Δk.

Hence the whole force pressing upwards is

$$-\Delta k - \frac{d\Pi'}{dx} k.$$

The mass is $\Delta . k$.

Therefore

$$\frac{d^2(x+X)}{dt^2} \text{ or } \frac{d^2X}{dt^2} = g\left(-1 - \frac{1}{\Delta} \cdot \frac{d\Pi'}{dx}\right).$$

Or since $\Pi' = H . \Delta'$,

$$\frac{d\Pi'}{dx} = H . \frac{d\Delta'}{dx};$$

and
$$\frac{d^2X}{dt^2} = -g - \frac{gH}{\Delta} \cdot \frac{d\Delta'}{dx}$$

$$= -g - \frac{gH}{\Delta} \cdot \frac{d}{dx}\left(\Delta - \Delta . \frac{dX}{dx}\right)$$

$$= -g - \frac{gH}{\Delta} \cdot \frac{d\Delta}{dx} + \frac{gH}{\Delta} \cdot \frac{d\Delta}{dx} \cdot \frac{dX}{dx} + gH . \frac{d^2X}{dx^2}.$$

But since $\Delta = \frac{\Pi}{H}$,

$$\frac{1}{\Delta} \cdot \frac{d\Delta}{dx} = \frac{1}{\Pi} \cdot \frac{d\Pi}{dx} = \frac{-1}{H}.$$

Also $$gH = a^2.$$

Hence $$\frac{d^2X}{dt^2} = -g + g - g\frac{dX}{dx} + a^2\frac{d^2X}{dx^2}.$$

Or, $$\frac{d^2X}{dt^2} + g\frac{dX}{dx} - a^2\frac{d^2X}{dx^2} = 0.$$

This equation cannot (with our present knowledge) be generally solved in a finite form.

The equation may be put into a form admitting of an apparent symmetrical solution in infinite series, by using u and v as the independent variables. By *Partial Differential Equations*, Article 35,

$$\frac{d^2X}{dt^2} - a^2\frac{d^2X}{dx^2} = 4a^2\frac{d^2X}{du\,.\,dv} = 4gH.\frac{d^2X}{du\,.\,dv}.$$

And $$\frac{dX}{dx} = \frac{dX}{du} \cdot \frac{du}{dx} + \frac{dX}{dv} \cdot \frac{dv}{dx} = \frac{dX}{du} - \frac{dX}{dv}.$$

The equation now becomes

$$\frac{d^2X}{du\,.\,dv} = \frac{1}{4H}\left(\frac{dX}{dv} - \frac{dX}{du}\right).$$

Neglecting the second side,

$$X = \phi(u) + \psi(v).$$

Substituting this value in the second side, and integrating

$$X = \phi(u) + \psi(v) + \frac{1}{4H}\int_u \cdot \int_v \cdot \left(\frac{dX}{dv} - \frac{dX}{du}\right)$$

$$= \phi(u) + \psi(v) + \frac{1}{4H}\left(\int_u X - \int_v X\right)$$

$$= \phi(u) + \psi(v) + \frac{1}{4H}\left(\int_u - \int_v\right)\{\phi(u) + \psi(v)\}.$$

Continuing this process we find

$$X = \phi(u) + \psi(v) + \Sigma\left[\left(\frac{1}{4H}\right)^p \cdot \left(\int_u - \int_v\right)^p \cdot \{\phi(u) + \psi(v)\}\right],$$

where p is to have the values 1, 2, 3, ... ∞.

We doubt, however, whether anything is really gained by this form.

56. *Investigation of the motion of a spherical divergent wave, from considerations of the movement of particles in a solid-angle-sector or pyramid.*

In Articles 44 and 50, the motions of symmetrical divergent waves, in space of two or of three dimensions, are treated by means of the Characteristic Function. But they may be treated by the more simple and direct method of considering the changes of volume and density, and the forces thence arising, in a solid-angle-sector or pyramid of very small angle. We will commence with a re-investigation of the problem of Article 50, the divergence of a wave in space of three dimensions.

Let the measure of the solid angle, estimated by the area of the transverse section of the pyramid at distance 1 from the origin, be s; let r be the distance of any particle in its undisturbed state, $r+T$ its distance at the time t. Between the transverse section at distance r and that at $r+\delta r$, the included volume is $s \,.\, r^2 \,.\, \delta r$. When r is changed to $r+T$, $r+\delta r$ is changed to

$$r+\delta r+T+\frac{dT}{dr}\,\delta r\,;$$

and between the transverse sections at those distances, the included volume is

$$s\,(r+T)^2\,.\left(\delta r+\frac{dT}{dr}\,\delta r\right).$$

The density Δ, in the neighbourhood of the disturbed particles whose original distance was r, is therefore

$$D\times\frac{s\,.\,r^2\delta r}{s\,(r+T)^2\,.\left(\delta r+\dfrac{dT}{dr}\,\delta r\right)}$$

$$=D\times\left(1-\frac{2T}{r}-\frac{dT}{dr}\right);$$

and the elastic pressure Π on a unit of surface is

$$=P\left(1-\frac{2T}{r}-\frac{dT}{dr}\right).$$

For the particles whose original distance was $r + r'$, the pressure Π' on a unit of surface

$$= \Pi + \frac{d\Pi}{dr} r'.$$

The mass of matter, between a unit of surface at one transverse section of the sector and a unit of surface at the other transverse section, is Dr'.

It might be thought at first that, in order to determine the movement of the truncated pyramid included between the original distances r and $r + r'$, we ought to compare the whole pressures at the two ends of the truncated pyramid. But on consideration it will be seen that, if we form a principal cylinder whose base is the smaller end, and produce it till it meets the larger end (as in Figure 11); and if we estimate the difference of pressures of those equal bases of the cylinder, and compare it with the distance between them; and if we also form parallel small cylinders, occupying with their bases the remaining part of the large end, and whose opposite extremities cut the inclined boundary; and if we estimate the resolved part of pressure on that inclined boundary in the direction of the cylinder's length; we shall find the difference of opposite forces to bear in all parts the same proportion to the length of the column of air which they are to move.

The acceleration therefore, or

$$\frac{g \times \text{difference of pressures}}{\text{mass moved}},$$

will be

$$-\frac{g}{Dr'}\cdot\frac{d\Pi}{dr}r' = -\frac{gP}{D}\cdot\frac{d}{dr}\left(1-\frac{2T}{r}-\frac{dT}{dr}\right)$$

$$= gH.\frac{d}{dr}\left(\frac{2T}{r}+\frac{dT}{dr}\right).$$

Making this $=\frac{d^2}{dt^2}(r+T)=\frac{d^2T}{dt^2}$,

we have

$$\frac{1}{a^2}\cdot\frac{d^2T}{dt^2}=\frac{d}{dr}\left(\frac{2T}{r}+\frac{dT}{dr}\right).$$

Let $T = Vr$;

then $$\frac{r}{a^2}\cdot\frac{d^2V}{dt^2}=4\frac{dV}{dr}+r\frac{d^2V}{dr^2},$$

or $$\frac{1}{a^2}\cdot\frac{d^2V}{dt^2}=\frac{4}{r}\cdot\frac{dV}{dr}+\frac{d^2V}{dr^2};$$

which being solved by the process of Article 41 gives

$$V=\frac{dH}{dv}\cdot\frac{1}{r^2}+\frac{H}{r^3},$$

(H being any function of v), and

$$T=\frac{dH}{dt}\cdot\frac{1}{r}+\frac{H}{r^2};$$

an expression equivalent to

$$-\frac{D'}{ar}-\frac{D}{ar^2},$$

found in Article 49.

(The same equation is solved by a different process, in the Author's *Partial Differential Equations*, Article 42.)

In the same manner we may proceed with such a problem as the following. To find the law of motion of a wave of air diverging simultaneously from all parts of the earth; supposing gravity to be inversely as the square of distance from the center, and the elastic pressure to vary as the nth power of the density (n being somewhat greater than 1, to make the atmosphere finite). It is necessary first to investigate the statical problem (as in last Article) and then to make the dynamical investigation (as above). An equation will be obtained in the form

$$\left\{A - \frac{(n-1)\,a^2}{r}\right\}\frac{d}{dr}\left(\frac{dT}{dr} + \frac{2T}{r}\right)$$

$$+ \frac{na^2}{r^2}\left(\frac{dT}{dr} + \frac{2T}{r}\right) + b^2\frac{d^2T}{dt^2} - \frac{4a^2T}{r^3} = 0;$$

a being the earth's radius, $b^2 = \dfrac{1}{g}$, $A = -nH + (n-1)\,a$.

(This equation was printed erroneously in the Author's *Partial Differential Equations*, Article 44.)

If this equation could be solved, it might give some information on the ultimate effect of radiations of various kinds from central bodies.

57. *Remarks on the increase due to the computed velocity and pressure in all the preceding results.*

Commencing with Article 32, we have, for convenience, systematically omitted the factor $n^2\theta^2$, in expressing the proportion of the change of elastic pressure to the change of density of the air. But this factor, or one of very nearly the same value, ought in all cases to be introduced. This will be seen on considering the characters of the several cases to which our theorems apply.

The necessity for employing the factor θ is explained in Article 14; the symbol being first introduced in Article 17 and first applied in Article 21. There can be no doubt on the necessity for using it in all cases.

The necessity for employing the factor n is explained in Articles 15 and 16; the symbol is first introduced and first applied in Articles 16 and 21. The reasoning in Article 15 shews that its existence depends entirely on the rapidity of change in the state of condensation of the air. The investigations in Articles 24, 25, 27, 30, 31, 32, 39, 44, 46, 48, 50, and in fact every investigation relating to sound, imply vibrations which go through their period several hundred times in one second, and for these the factor n or 1·2 must indubitably be used. The effect of it is that the velocity of transmission of the wave instead of being a or $\sqrt{gH'}$

will be $1\cdot 2 \times \sqrt{gH'}$. The investigation of Article 51, or rather its application to the pendulum-vibration in Article 54, implies vibration in which the change from least to greatest density and *vice versâ* occupies one second. It is certain, from the experiments mentioned in Article 15, that a great effect of the kind sought takes place in that duration of time, but it seems doubtful whether the whole effect corresponding to rapid vibrations will be produced. If the whole effect is really produced in that case, then, since the object of the investigation was merely to ascertain pressures, these pressures will be increased in the proportion of $1 : N$ or $1 : n^2$ (see Article 16); and in Article 54 the "proportion of the pressure in the direction of motion produced by air-elasticity to the pressure produced by gravity" will be $-\frac{n^2}{2} \cdot \frac{\text{weight of displaced air}}{\text{weight of pendulum-bob}}$, or

$$-0\cdot 72 \times \frac{\text{weight of displaced air}}{\text{weight of pendulum-bob}};$$

and the final result of that article will be

$$1\cdot 72 \times \frac{\text{specific gravity of air}}{\text{specific gravity of pendulum-bob}}.$$

58. *On the combination of similar waves travelling in opposite directions; and on stationary waves.*

If, in the motion of waves of air through a cylinder, or in the motion of a plane wave of air in space, (Articles 24, 25, 27, 39), the expression for displacement of a particle consists of two functions expressing wave-motion in opposite directions, as

$$\phi(n\theta a \,.\, t - x) + \psi(n\theta a \,.\, t + x),$$

and if the two functions be trigonometric, similar, and with equal coefficients, as

$$b \,.\, \sin(n\theta af \,.\, t - fx + c) + b \,.\, \sin(n\theta af \,.\, t + fx + e),$$

their sum takes the form

$$X = 2b \,.\, \sin\left(n\theta af \,.\, t + \frac{c+e}{2}\right) \times \cos\left(fx + \frac{e-c}{2}\right).$$

Here the appearance of a travelling wave is lost entirely. The function presents none of the characteristics described in Articles 24 and 25. Yet every particle of air, with only critical exceptions, has a vibratory motion.

If we consider a single particle, that is, if we make x constant, that particle has a vibration whose coefficient is $2b \,.\, \cos\left(fx + \frac{e-c}{2}\right)$, and whose law of oscillation is $\sin\left(n\theta af \,.\, t + \frac{c+e}{2}\right)$, going through all its changes in the time $\frac{2\pi}{n\theta af}$. The magnitude of vibrations is different for different particles; but the beginning and end of vibration occur at the same time for all the particles. The coefficient vanishes, that is, the particle has no oscillation, where

$$fx + \frac{e-c}{2} = \frac{\pi}{2}, \ \frac{3\pi}{2}, \ \frac{5\pi}{2}, \ \&c.$$

If we consider all the particles at one moment, that is, if we make t constant, the displacement of different particles follows the law

$$\cos\left(fx+\frac{e-c}{2}\right).$$

It has one sign between

$$fx+\frac{e-c}{2}=\frac{\pi}{2} \text{ and } =\frac{3\pi}{2},$$

the opposite sign between

$$fx+\frac{e-c}{2}=\frac{3\pi}{2} \text{ and } =\frac{5\pi}{2},$$

the first sign between

$$fx+\frac{e-c}{2}=\frac{5\pi}{2} \text{ and } =\frac{7\pi}{2},$$

&c.

There is constant quiescence where

$$fx+\frac{e-c}{2}=\frac{\pi}{2}, \text{ or } \frac{3\pi}{2} \text{ or } =\frac{5\pi}{2},$$

&c.

The maximum value of displacement is

$$\pm 2b\,.\sin\left(n\theta af\,.\,t+\frac{c+e}{2}\right),$$

always occurring, whatever be the time, where

$$fx+\frac{e-c}{2}=0,$$

or $=\pi$, or $=2\pi$, &c.

From the last mentioned characteristics, this law of disturbance has received the name of *the stationary wave.*

The variable part of the elastic force of the air (Article 21) is $-D.\frac{dX}{dx}$,

or $$2bfD.\sin\left(n\theta af.t+\frac{c+e}{2}\right).\sin\left(fx+\frac{e-c}{2}\right).$$

This, for any one particle, has for coefficient

$$2bfD.\sin\left(fx+\frac{e-c}{2}\right).$$

This coefficient is greatest where

$$fx+\frac{e-c}{2}=\frac{\pi}{2},\ \frac{3\pi}{2},\ \&c.,$$

and vanishes where

$$fx+\frac{e-c}{2}=0,\ \pi,\ 2\pi,\ \&c.$$

On comparing this with the preceding statements, it appears that those points of the air which have no displacement have the greatest change of density, and those which have the greatest displacement have no change of density.

The reader will at once perceive that the theory of the echo, in Article 41, applies to this case; supposing $\alpha=\frac{\pi}{2},\ \beta=\frac{\pi}{2},\ \gamma=0.$

Stationary waves are formed, in like manner, by the combination of converging and diverging waves in the cases supposed in Articles 44, 46, 50, 51. The reader will have no difficulty in verifying this. It will be necessary to distinguish carefully the signs of $\frac{dR}{dr}$ as derived through u and as derived through v.

59. *Deficiences still existing in the mathematical theory of atmospheric vibrations, as applied to important cases occurring in practice.*

The first deficiency to which we shall advert is in the general treatment of the reflection of waves of air. We have seen in Article 41, &c. that the reflection of ordinary plane waves of air at a plane surface is treated theoretically without difficulty; and if we should use a similar process for such plane waves as occur in Article 42 (the formulæ of that Article being so altered as to represent two directions of motion of wave inclined to the axis z, in order to exhibit the wave in the generality of inclination), or for such diverging waves as occur in Articles 44, 46, 50, and 51 (with due alteration for representing the places of the two centers of divergence and the two directions of oscillation), we should find no difficulty, provided we assume that the surface of reflection (that is, the surface along which the motions of the particles produced by the combination of two waves are at all times parallel to the portions of surface which they touch) is a plane.

But if we assume that surface to be curved, we meet with difficulties. It might be supposed that, with a parabolic surface, the movements of particles, produced by a spherical wave diverging from the focus, and by a plane wave moving in the direction of the axis, would be so related as to shew that, by reflection at the parabolic surface, one of these waves would be the consequence of the other. This, however, has not been proved algebraically, and appears to be doubtful. In like manner, it might be supposed that, with a prolate spheroidal surface, the movements of particles in waves diverging from one focus and converging to the other focus would possess the relation proper for reflection; but this is equally in doubt. And these doubts apply to reflection at a curved surface generally.

The second deficiency is in the investigation of the motions of the particles at the junction of two containing vessels. Suppose, for instance, we consider a large tube stopped at one end and communicating at the other end with the open air. There is no difficulty in understanding that there may be a stationary wave in the tube (the stopped end being one of the points of vanishment of motion in Article 58), and that there may be a stationary divergent wave in the open air. But, if so, where will be the next surface of vanishment of motion? or that of vanishment of variability of pressure? Theory has not yet answered these questions.

We commend these problems to the attention of the student.

SECTION V.

TRANSMISSION OF WAVES OF SONIFEROUS VIBRATIONS THROUGH DIFFERENT GASES, SOLIDS, AND FLUIDS.

60. *Velocity of waves through gases.*

The investigation of Article 21 applies in the same manner to all gases as to atmospheric air, excepting that we are not so well acquainted with the effects of change of temperature and sudden contraction or expansion. Omitting these, that is, putting 1 for $n\theta$, we find as in Article 21, that about the particle whose original ordinate was x, the density of the gas is represented by $D - D\frac{dX}{dx}$; and by Boyle's Law, Article 10, which is found to apply to all gases, the elastic pressure of the gas about that point is therefore

$$K\left(D - D\frac{dX}{dx}\right),$$

K being a constant of whose value we shall speak very soon. Consequently, the elastic pressure about the point whose original ordinate was $x + k$ is

$$K\left(D - D\frac{dX}{dx} - D\frac{d^2X}{dx^2}k\right);$$

and the excess of the former mentioned pressure, tending to move the included mass of gas forwards, is

$KD\frac{d^2X}{dx^2}k$; the mass of gas to be moved is Dk; and therefore we have the differential equation

$$\frac{d^2(x+X)}{dt^2} \text{ or } \frac{d^2X}{dt^2} = gK\frac{d^2X}{dx^2};$$

of which the solution is

$$X = \phi(\sqrt{gK}.t - x) + \psi(\sqrt{gK}.t + x);$$

indicating, as in Articles 24 and 25, waves whose velocity is $\sqrt{gK}$.

In order to explain what is meant by K, let us suppose that the gas is contained in vessels which are not rendered, by the inclosed gas, liable to any strain either of bursting or of contraction, and therefore that the gas exerts the same elastic force as the external air. In this state, let the specific gravity of the gas be $= G \times$ specific gravity of the external air; or, using our own language, let the density of the gas be $G \times$ density of air. Now by Articles 8 and 13, the elastic pressure of the air is able to support the weight of a column of similar air whose height is H; or, in Article 13, the elastic pressure of air $= H \times$ density of air. But we have just supposed that the elastic pressure of the gas is equal to that of the air, and that the density of the air $= \frac{1}{G} \times$ density of gas. This equation therefore becomes

$$\text{Elastic pressure of gas} = \frac{H}{G} \times \text{density of gas.}$$

Consequently our factor K is $= \frac{H}{G}$; and the velocity of waves is $\sqrt{\frac{gH}{G}}$. In air, treated in the same manner, the velocity of waves is $\sqrt{gH}$. Therefore the velocity of waves in gas is $\frac{1}{\sqrt{G}}$ × velocity of waves in air: where G is the specific gravity of the gas, referred to air at the same pressure, as the standard of specific gravity.

61. *Velocity of waves through solid bodies.*

The fact of the transmission of vibrations through a solid body, and the calculation of the velocity of the wave, rest on the assumption that the particles of the body are so connected, that compressive force from some external cause is necessary to make the particles approach nearer together, and that extensional force is necessary to separate the particles more widely; and that the effects so produced will be proportional to the forces producing them. In the extension of a longitudinal bar of metal, we may thus represent the law: the weight which, if applied for extension, will increase the intervals between particles in the proportion $1:1+z$, or which, if applied for compression, will diminish the intervals between particles in the proportion $1:1-z$, must be the weight of a similar bar of the same metal whose length is $L.z$ (where L is a given length or modulus, peculiar to the metal). Adopting for malleable iron as engineer's data (rather uncertain), that a weight of 1 lb. or 3·6 cubic inches of iron will extend an iron bar whose section is 1 square inch by $\frac{1}{24000000}$ of its length without injuring its elasticity, the weight of a length Lz inches of a similar bar, that is, the weight of Lz cubic inches of iron, will extend it by the fractional part

$$\frac{Lz}{3 \cdot 6} \times \frac{1}{24000000}.$$

Making this $= z$, we have $L = 3{\cdot}6 \times 24000000$ inches $= 86400000$ inches $= 7200000$ feet. Then an extension of the space occupied by given particles in the proportion $1 : 1 + z$ implies that they are subjected to an extensional force equal to the weight of a similar bar whose length is $z \times L$, or (for iron) $z \times 7200000$ feet, and consequently that they pull with that force on the connexions which extend them; and similarly, *mutatis mutandis*, for a contraction of the space.

If now the particle, whose distance from origin was originally x, is disturbed through X, the particle whose distance was $x + h$ will be disturbed through

$$X + \frac{dX}{dx} h;$$

the space occupied by particles, which was originally h, will now be $h + \frac{dX}{dx} h$; the space is increased by the fractional part $\frac{dX}{dx}$, which is to be put for z in the last paragraph; and the particles which were originally in the position x are now pulling those on both sides of them with a force equal to the weight of a similar bar whose length is $\frac{dX}{dx} \times L$. They are therefore pulling the particles in front of them backwards, with that force. For the particles which were originally in the position $x + k$, there is a force pulling the particles in rear of them forwards, with the force

$$\left(\frac{dX}{dx} + \frac{d^2X}{dx^2}k\right) \times L.$$

The excess of the latter gives the force really pulling the intervening particles forward

$$= \frac{d^2X}{dx^2}k \times L.$$

The mass of matter intervening, estimated in the same manner, is k. Hence the acceleration forwards is

$$\frac{d^2X}{dx^2} \times gL;$$

and the equation is

$$\frac{d^2(x+X)}{dt^2} \text{ or } \frac{d^2X}{dt^2} = gL \times \frac{d^2X}{dx^2};$$

the solution of which is

$$X = \phi(t\sqrt{gL} - x) + \psi(t\sqrt{gL} + x);$$

and the velocity of the waves is $\sqrt{gL}$.

With the data above given for iron, this velocity is 15203 feet per second. This value is larger than that for any other metal.

62. *Velocity of waves of sound through fluids.*

The theory of the transmission of vibrations through fluids is embarrassed with a complication from which that of transmission through solids is free. The ordinary laws of equality of pressure in all directions apply, apparently, in the same manner to those sudden shocks which are distributed by pulses similar to those of sound, as to those slower communications of motion which are transmitted by visible waves. We have remarked, when in a barge on the sea at some distance from the vertical of the spot where a large quantity of gunpowder was fired at about 60 feet depth, that a sudden shock was felt upwards at the bottom of the barge long before there was the smallest sign of an ordinary wave. Here the shock had been communicated by molecular transmission in the same manner as through an iron bar, but with this difference of dispersion, that it had diverged through a solid angle. After a wide limit of space, remarking that the depth of a sea or lake is usually a very small fraction of its horizontal extension, it seems probable that the waves of vibrations will extend as confined between two horizontal planes. Thus, for a distance comparable to the depth of the water, the investigation of Articles 50 and 56 would nearly apply; for greater distances, that of Article 44 would appear to correspond better to the circumstances. In either case, however, as appears from Article 50, the velocity of the wave will not sensibly differ from that of a wave transmitted longitudinally through a uniform cylinder. The amplitude of vibrations at a distant point will be diminished, but much less under the hypothesis that the waves diverge as between two parallel planes, than if the waves diverged, through their whole course, into three dimensions.

Section VI.

EXPERIMENTS ON THE VELOCITY OF SOUND, AND ON THE PRESSURE ACCOMPANYING ATMOSPHERIC WAVES; AND COMPARISON OF THE EXPERIMENTAL RESULTS WITH THE RESULTS OF THEORY.

63. *Recapitulation of the theoretical results for the velocity of Sound in Air.*

It has been found, in Article 24, that the velocity of sound in a cylinder is $n\theta \times 916{\cdot}2722$ feet per second; where n is a constant (Articles 15 and 16) depending on the increase or diminution of the elastic force of the air produced by sudden compression or expansion, to which we have assigned the probable value 1·2; and where θ is a factor depending on the temperature of the air during the experiment, and represented by $\sqrt{\Theta}$ or

$$\sqrt{\frac{450 + \text{reading of Fahrenheit's thermometer}}{482}}$$

(Articles 14 and 17). Converting the formula into numbers, we have the following table of the theoretical velocities of sound;

Theoretical Velocity of Sound, at different temperatures of the air as indicated by Fahrenheit's Thermometer.

Temp.	Velocity.	Temp.	Velocity.
−40°	f. 1014·1	32°	f. 1099·5
−30	1026·4	40	1108·6
−20	1038·5	50	1119·9
−10	1050·5	60	1131·1
0	1062·4	70	1142·2
10	1074·1	80	1153·0
20	1085·7	90	1163·8
30	1097·2	100	1174·5

The only source of uncertainty in these numbers is the uncertainty on the value of n, (see Article 16). The numerical coefficients given by different theorists are sensibly different, and we are yet ignorant whether the value of n depends on the temperature of the undisturbed air.

It appears from Article 39 that the velocity of a plane wave of air is the same as that of a wave in a cylinder: and it appears from Article 50 that the velocity of a divergent wave is sensibly the same as that of a wave in a cylinder.

64. *Methods used for determining the velocity of Sound.*

In the greatest part of the experiments, the observations have been those of the flash and the report of a distant cannon. The flash, and the first disturbance of air by the emission of gas, occur so nearly or exactly at the same instant, that no sensible error arises from the difference in the nature of these two phenomena. The same observer observes both phenomena with the same watch or clock; and, if the distance of the gun be several miles, there is ample time for the observer to write down the observation of the flash before preparing himself for the observation of the sound. All these circumstances are very advantageous. The gun is usually pointed towards the observer, and it seems probable that this circumstance may slightly accelerate the pulse of air in the beginning of its course, but possibly by a few feet only, corresponding to an imperceptible error of time.

But there is a physiological circumstance, the effects of which have hitherto escaped notice, but which probably produces a sensible error; it is, that two different senses (sight and hearing) are employed in the observation of the two phenomena, and we are not certain that impressions are received on them with equal speed. Indeed we believe that the perception of sound is slower by a measurable quantity, perhaps $0^s \cdot 2$, than the perception of light; and this may affect the result with an error amounting to some hundreds of feet.

We should much prefer a plan of observation in which two observers observed, in the same manner, the time of the sound passing two stations. By using signals given reciprocally from two stations beyond both the observing stations, it will be easy to obtain a result for the time of passage of the sound, independent of the habits of each observer, independent of the difference of the indications of their time-keepers, and independent of the velocity of the wind. (The reader will verify this without difficulty, by putting algebraical symbols for the different elements just mentioned; when it will be found that, on taking the mean of the two apparent times occupied by the passage of sound, according as the gun at first station or at second station is used, those elements disappear.) A process of this kind is employed in the measurement of higher velocities, as the velocity of the galvanic current in a telegraph-wire.

Difficulties have sometimes been experienced, by persons not familiar with astronomical practices, in the estimation of fractions of a second of time. To avoid these, a timepiece was employed in the Dutch experiments to be mentioned below (perhaps, on the whole, the best which had been made before those of M. Regnault) in which the motion, being regulated by a pendulum revolving in a conical form, was free from the jerks of a common clock, and the index could be stopped at any fraction of a second.

A most elaborate series of experiments by M. V. Regnault is published in the *Mémoires de l'Académie des Sciences*, tome XXXVII., occupying 575 pages. The most important were made in tubes prepared for conveyance of gas and water in the neighbourhood of Paris: these tubes varied in diameter from 0·108 mètre to 1·10 mètre, and in length from 961 to 4886 mètres. The general principle in all was, to cover the near end of the tube with a firm plate (excepting in some early experiments), in which was a hole through which a pistol barrel was thrust; and a charge of powder, or sometimes a large percussion cap, was fired. The distant end of the tube was covered with a sheet of caoutchouc, which was made to tremble by the shock of the air-wave: sometimes it produced a reflexion to the firm plate, and from it to the caoutchouc again, &c. The pistol-explosion broke a galvanic circuit, and the trembling of the caoutchouc restored it: and these galvanic effects were registered upon a revolving barrel, on which were also registered the beats of a clock and the vibrations of a tuning-fork. In some experiments, laminæ of caoutchouc were applied to apertures in the sides of the tube at different distances. Finally, experiments were made in the same way without tubes, using the explosion of a heavy cannon. Experiments were also made on the velocity of sound through air of different densities, and through various gases. These, we believe, are the only experiments in which there has been no reference to human nerves.

65. *Statement of the principal modern results for the velocity of Sound.*

We confine ourselves, in the following table, to the most trustworthy experiments made since the year 1820.

Authority.	Experimenter.	Distance in feet.	Velocity in feet.	Temp. Fahren.
Phil. Trans. 1823	Goldingham	29547	1089·9	32°
		13932	1079·9	32
Conn. des Temps, 1825	Arago and others	61064	1086·1	32
Vienna Jahrbuch, Vol. VII.	Myrbach and Stamfer	32615	1092·1	32
Phil. Trans. 1824	Moll and Van Beek	57839	1089·4	32
Camb. Trans. Vol. II.	Gregory	13440	1097	33
			1085·8	64
		9874	1117	66
Parry's 3rd Voyage	Parry and Forster	12893	1014·4	−38·5
			1010·3	−37·5
			1029·0	−37·0
			1021·0	−24·5
			1026·6	−21·5
			1039·3	−18·0
			1037·3	− 9·0
			1040·5	− 7·0
			1098·3	+33·5
			1118·1	35
Mem. de l'Académie, tome 37	Regnault	Various	1085·0	32

The best of these experiments give results somewhat larger than those in our theoretical table, Article 63; some, however, give results rather smaller. It seems not impossible that we ought to have taken for n a value very slightly exceeding 1·2, and also that there is some uncertainty in the experiments.

It is to be remarked that, except in M. Regnault's experiments, there are no observations sufficient to enable us to apply a correction for the effect of moisture in the air.

M. Regnault's experiments shewed clearly that the greatest disturbance produced by a violent solitary wave travelled rather more rapidly than that of a feeble wave: as may be inferred from the theory of Article 33.

We ought not here to pass over a curious remark of Captain Parry and Mr Fisher, in observations made in Captain Parry's second voyage. The stations were so near that the human voice could be heard; and the remark was, that the officer's word of command "fire," was heard about one beat of the chronometer (or $\frac{2}{5}$ of a second of time, we believe) *after* the report of the gun. The instance is quite singular. Mr Earnshaw has supposed it possible that the phenomenon has some relation to that acceleration of the wave which occurs when the displacement of particles is very large (see Articles 32, 33, 34). But we cannot imagine that the acceleration could ever amount to a space of 200 feet; and in any case we imagine that when the accelerated large disturbance came up with the small disturbance, the two disturbances would be merged into one.

It appears to us more probable that the phenomenon is physiological. We have often remarked that, when the report of a gun or any other violent and sudden noise is heard, it is preceded by the perception of a shock through the bodily frame, the interval in time being a large fraction of a second. From the voice, there would be no sensible shock: but the shock from the cannon-explosion might be sensible, and might precede the auditory perception of the report by a time sufficiently long to present itself to the observer's mind before the auditory perception of the voice.

66. *Comparison, with theory, of the observed pressure accompanying an atmospheric wave: the pendulum.*

The only experiment which is sufficiently delicate to give a measure of the pressure of a wave is the observation of its influence on the movement of a pendulum whose bob is a sphere. And the reason why this experiment is so delicate is, that the effect of the pressure of the wave is to lengthen the time of every vibration; and though its effect on the time of a single vibration would be undiscoverable, yet its aggregate effect on the total time of a great number of successive vibrations is very conspicuous, and very little doubt exists as to the accuracy of the results found from it.

Referring to books on Mechanics for descriptions of the methods by which the length of a pendulum to its center of oscillation is accurately measured and its time of vibration is accurately observed, it is only necessary here to remark: that for different pendulums swinging in a vacuum, whatever be the materials of which they are composed, an invariable relation exists (at a given locality) between the length of the pendulum and its time of vibration: but that, in the far greater number of experiments, the vibrations have been observed not in a vacuum but in air; and that therefore a numerical correction to the observed time of vibration of each is necessary, in order to produce the time of vibration which would have been observed in vacuum. In the reduction of all the earlier pendulum-observations, the numerical correction was computed on the supposition that the sole effect of the air was, to diminish the active weight of the bob by an absolute quantity equal to the weight of the displaced air, and therefore to diminish the effect of gravity by a proportional part represented by a fraction, whose numerator is the specific gravity of air and whose denominator is the specific gravity of the pendulum-bob.

But it was found in the present century (in the first instance, we believe, by Bessel) that when different pendulums composed of different metals were treated in this way, they gave discordant results: the relation between the pendulum's length and its corrected time of vibration did not hold uniformly. In all cases it was necessary to apply a larger numerical correction than that given by the rule which we have just stated.

Experiments therefore were made, principally by

M. Bessel in Prussia, and Mr F. Baily in England, in order to discover how much the original correction ought to be increased. (Observations, not applying to the spherical form, were also made by Captain Sabine.) In some instances, spheres of different substances were used in the same state of air; in other instances, the spheres were not varied, but the density of the air was varied by conducting the experiments in a close case from which the air was partially exhausted by an air-pump. The results were the following;

Bessel, by comparing the vibrations of a sphere of brass and a sphere of ivory, in common air, found that the old correction ought to be multiplied by 1·95.

Baily, by comparing the vibrations of a sphere in common air with the vibrations of the same sphere on the same mounting in vacuum, and applying the same process to different spheres, found for the factor: with spheres 1½ inch in diameter, platinum 1·881, lead 1·871, brass 1·834, ivory 1·872; with spheres 2 inches in diameter, lead 1·738, brass 1·751, ivory 1·755.

Now, in Article 54, corrected by the considerations cited in the latter part of Article 57, we have found, as the theoretical result of considering the spherical wave with oscillating centre, that the old correction ought to be multiplied by 1·72.

We consider the agreement of the observed and the theoretical results as being as good as, under all the experimental circumstances (especially with the limitation of the surrounding space of air), could be expected.

67. *Remarks on whispering-galleries.*

We have called attention in Article 59 to the difficulty in the theoretical treatment of reflexions of waves of air at curved surfaces. The effect of what is usually called a "whispering-gallery" is included in this case; and we can therefore offer only a popular and imperfect account of it. It will be remarked that the theory of reflexion of waves of air from a plane surface is perfect; and in a hemispherical dome of very large dimensions the curvature is small, and we may expect the theory of plane surface to apply with some degree of approximation. Suppose then a series of sound-waves to issue in a divergent form through a not very large angle from the speaker's mouth. Different parts of this series, following different radii of divergence, will meet the slightly-curved dome-surface at different distances: but each part will then be reflected in such a direction that it again meets the surface at successive points of contact after successive equal chords; and, for each part of the series, all these chords are in such a plane that, for each part of the series, the reflexions tend towards the point of the hemisphere opposite to the speaker; to which point there is consequently a general convergence of reflexion-paths. Moreover, though the lengths of the chords are different, yet for each part the polygonal sum of chords differs little (perhaps a foot or two) from the curved line on the dome-surface, and therefore the different waves from different parts meet at that opposite point in nearly the same phase. Hence they produce by their union a sound-wave of considerable intensity.

It is proper to remark that the peculiar sound of a whisper is not required for exhibition of the effect; low articulate sounds of musical character are reflected in the "whispering-gallery" with the same perfectness as a genuine whisper.

68. *Experiments on the velocity of Sound through gases.*

It is impossible for us to form an atmosphere of hydrogen gas or of carbonic acid extending several miles, and therefore it is impossible for us to experiment on the velocity of sound through gas in the same way as through air. To explain the process which has been successfully used, we must here anticipate the results of a subsequent section. It must be understood then that when an organ-pipe is sounded in the usual way, the frequency of sound-waves which it produces depends upon the time occupied by a wave's travelling from one end of the pipe to the other (or, in certain cases, travelling twice the length of the pipe); and it must also be understood that every definite frequency of sound-waves produces a definite musical note to the ear. Consequently, with a given organ-pipe, the musical note produced will depend on the velocity of the wave's travel; and the accurate observation of the musical note will give accurate information on that velocity. It is only necessary therefore to inclose an organ-pipe in an atmosphere of the gas upon which it is desired to experiment, and to adapt to it apparatus for blowing the gas in the same manner in which air is blown for the ordinary sounds of the organ-pipe, and to remark the note which it produces; the relation of this note to the note which the same pipe produces in air, interpreted with reference to the theory of musical tones, which we shall explain in a subsequent section, gives the proportion of the frequencies of sound-waves in the pipe, and the proportion of the velocities of the wave's progress, in the gas and in air.

Thus the experimental numbers in the following table have been obtained (Dulong, *Mémoires de l'Institut,* Tome x.). Instead of giving the actual velocity of sound in each gas, it has appeared more convenient to give the proportion of each velocity to the velocity in

Name of Gas.	Proportion of specific gravity to that of air.	Theoretical proportion of sound-velocity in gas to that in air.	Observed proportion of sound-velocity in gas to that in air.
Oxygen gas	1·1026	0·9523	0·9525
Hydrogen gas	0·0688	3·8125	3·8123
Carbonic acid	1·524	0·8100	0·7855
Oxide of carbon	0·974	1·0133	1·0132
Oxide of azote	1·527	0·8092	0·7865
Olefiant gas	0·981	1·0096	0·9439

air at the same temperature. The theoretical proportions of velocities are computed by the formula of Article 60.

The defect in the observed velocity in carbonic acid, oxide of azote, and olefiant gas, indicates a value of *n* (Articles 16, 21), smaller for those gases than for atmospheric air. This circumstance is connected with a chemical theory of "specific heat," for which we refer the reader to treatises on Modern Chemistry.

M. Regnault found from direct experiments in tubes,

Hydrogen	3·801
Carbonic Acid	0·7848 0·8009
Oxide of Azote	0·8007
Ammoniacal gas sp. grav. 0·596	1·2279

69. *Experiments on the velocity of Sound through solid bodies.*

It is easy to perceive the difference in the velocities of sound, as transmitted by the air, or as transmitted by metals (where the portions of metal are united by solder, &c. so as to form a continuous piece of great length, or where their parts are forced into firm contact by considerable tension). We have remarked, for instance, that a strong chain, lying upon a long and steep incline of a railway, transmits sound well. If the chain is struck at one end, and an observer at the other end applies his ear to it, he will perceive two sounds; the first conveyed by the metal, the second (which travels more slowly) transmitted by the air. It was in this manner that Biot (*Traité de Physique*) made experiments on a length of 951 mètres of cast iron pipes, from which he concluded that the velocity of sound in iron is 10·5 × velocity of sound in air; and Wertheim (Poggendorf, *Ergänzungsheft* III.), using 4067·2 mètres of telegraph wires, found a velocity of 3485 mètres per second, which differs little from Biot's.

Attempts however have been made to measure the velocity by experiments on a small scale (see Wertheim, *Annales de Chimie*, 3me série, Tome XII.). Without going into details of complicated apparatus, we shall state that, by reference to musical note (as will be mentioned in a subsequent Section), the rapidity of vibration of a tuning-fork is known; and that this can be exhibited by scratches which it makes on a glass surface moving under it, slightly covered with lamp-black. Transversal vibrations of a given bar, treated in the same manner, were made comparable with the tuning-fork-vibrations; and longitudinal vibrations, in like manner, were made comparable with the transversal vibrations. The time occupied by a longitudinal vibration was held, as that of air in an organ-pipe (hereafter to be mentioned), to be the time in which a wave passed through the double length of the bar. Thus the velocities, as compared with that in air, were found for different metals; the highest being that of iron, 15·108; the lowest that of lead, 3·974. Experiments were made at the same time on the extensibility of the metals; these, interpreted by the theory of Article 61, gave for velocities, in iron, 15·472, in lead, 3·561; differing little from the former.

A more remarkable method, however, has lately been introduced (see Kundt, Poggendorf's *Annalen*, Vol. 127). If a bar of metal, &c. be chafed, it is put into longitudinal vibration; and if its end carry a light piston, which nearly fits without touching the inside of a glass tube, vibrations of the same period will be excited in the air within the tube, and the lengths of the waves of these vibrations may be made visible by scattering a very light dry dust in the interior of the tube, which dust collects in little heaps in those parts of the tube where the air has no motion (Article 58), and the corresponding length of the wave of air is therefore known. In this manner, the length of wave in the metal, &c. (which is the double length of the bar), is immediately comparable with the length of wave of the same period in air; and when the periods are equal, the velocities of the transmission of waves are in the same proportion as the lengths of the waves (Article 30). Thus the following proportions of the velocities of sound-waves to the velocities in air were found:—

in steel, 15·34;
in glass, 15·25;
in copper, 11·96;
in brass, 10·87.

The velocity of sound in a stretched wire, confined at its ends, may be found by chafing it longitudinally and observing the musical tone which it produces; the number of waves corresponding to that tone being known (Article 94, below), the double length of the wire must be multiplied by the number of waves.

The velocity of sound through wood has been found in the same way. Along the fibre, it varies from 10900 to 15400 feet per second; transversally across the rings, from 4400 to 6000; and transversally along the rings, from 2600 to 4600 (Wertheim, *Mémoires*).

70. *Experiments on the velocity of Sound through fluids.*

A most important series of experiments was made by MM. Colladon and Sturm, for ascertaining by direct observation the time occupied by sound in passing through the water of the lake of Geneva (*Annales de Chimie,* Tome 36). The method adopted was, to suspend a large bell in the water, and to strike it with a hammer; at the place of observation, a tube was inserted in the water, having a large spoon-shaped orifice at its lower end, turned towards the origin of sound, and having a conical form at the upper end, terminating in a small hole to which the observer's ear was applied. The sound of a bell, weighing 500 kilogrammes (½ ton), sunk 3 mètres (10 feet) deep in the water, and struck by a man with a hammer weighing 10 kilogrammes (22 lbs.), was heard very well at the distance of 35000 mètres (nearly 22 miles). But the experiments which they were enabled to carry to the greatest extent were made with a bell only 7 decimètres high, suspended 1 mètre deep in the water, with a striking apparatus so arranged that at the instant of striking the bell it fired some gunpowder; the observer was stationed at a distance of 13487 mètres (44250 feet, or more than 8 miles). The velocity found was 1435 mètres or 4708 feet per second. The temperature of the water was 8°·1 centigrade.

Experiments made by the same philosophers on the compressibility of water gave, for the compression produced by the weight of one atmosphere, 49·5 millionth parts of the whole. From this, using formulæ similar to those of Article 62, they inferred a theoretical velocity 1428 mètres, agreeing well with that which was observed.

Wertheim (*Annales de Chimie*, 3rd series, Vol. XXIII.) has attempted an experimental determination of the velocity of sound in water by the same method which we have described above (Article 68) as applied to various gases, namely, by immersing an organ-pipe in water, and forcing the water through it in the same manner as air; a sort of musical tone was produced, sufficiently good to have its pitch recognized. The velocities found for water of the Seine varied from 1173 mètres to 1480 mètres per second; all much inferior to those found by the direct experiment. To reconcile them, Wertheim supposed that the velocities in a column of water and in an unlimited space of water are not the same; an idea which we do not accept. Among possible causes of the difference, we might suggest the yielding of the sides of the tube when pressed by the vibrations of a dense liquid; the yielding would be insensibly small when the vibrating mass was air or any other gas.

We are not aware that any attempt has been made to determine by measure the *amplitudes* (see Article 30) of the vibrations of particles of air in a Wave of Sound. It appears probable that they are extremely small. The sound of an ordinary tuning-fork (Article 71) held close to the ear is very loud, but its vibrations are invisible to the eye. They have been made sensible, in Professor Tyndall's experiments, by reflexion of light.

SECTION VII.

ON MUSICAL SOUNDS, AND THE MANNER OF PRODUCING THEM.

71. *The characteristic of the disturbance of air which produces on the ear the sensation of a musical note is, the repetition of similar disturbances at equal intervals of time; the greater rapidity of repetition producing a note of higher pitch. The Tuning-fork, the Siren, the Reed, the Monochord.*

Where movement of the air is produced by movement or mechanical shock of solid bodies, or even where it is produced by gaseous movements similar in their nature to those of solid bodies, we are able to trace their soniferous effects with great precision. And thus we soon arrive at the difference between *noise* and *musical sound.* If a cart-load of stones is emptied upon a hard road, it produces noise. Thunder is a noise. An axe or a hammer striking a tree produces a noise. The firing of a gun (that is, the sudden development of gas from a firm tube) produces a noise. In all these cases, there is either a single shock of the air, or (as in thunder) an irregular repetition, at sensible intervals, of single shocks, each of which may be considered as generating a solitary wave, such as is treated in Article 31.

But if any repetition be made, either by periodic shocks upon a hard substance whose agitations produce agitations in the air, or by periodic interruption or modification of a current of air,—provided that such shocks or interruptions be similar in character and uniform in interval of time, and provided also that the frequency be included within certain very wide limits (from about 30 in a second of time to about 10000 in a second, or even through a wider extent),—then a musical note is produced. The more rapid is the succession of shocks, &c. the higher is the pitch of the note.

It is said that Galileo first remarked the production of a musical note by the repetition of unmusical shocks, on passing a pen rapidly upon the milled edge of a coin, which made a small snap at every roughness. The experiment has often been repeated by snaps of a quill upon the teeth of a wheel in rapid motion. But the instruments which give the most satisfactory evidence of the truth of the assertion are the following:

(1) The tuning-fork, described in Article 49. In every elastic metal, when it is disturbed from a form of rest, in a definite manner, the force tending to restore it to its original state is exactly proportional to the extent of displacement; a state of things represented by the differential equation $\frac{d^2z}{dt^2} = -A^2 . z$; and then the law of displacement as connected with the time is accurately a law of sines; the solution of the equation being

$$z = B . \sin(At + C).$$

Here it is certain that the vibrations of air which are communicated to the ear follow the simple law of sines, but the rapidity of the vibrations is not ascertained. The note of the tuning-fork strikes the ear as being remarkably pure.

(2) The Siren. Suppose a flat disk, pierced with a great number of holes at equal distances round its circumference, to be so placed that the nozzle of a bellows can blow directly through any of the holes when by the rotation of the disk the hole is brought under the nozzle; and suppose that the disk is made, by clock-work, to rotate with a great speed which is registered by the clock-dials. Here we have a current of air interrupted very frequently, and at proper speed a powerful and sweet musical note is produced. The power is increased if, instead of having a single outlet of air, a plate similar to the rotating disk and having the same number of holes is firmly fixed near it, and the air is driven through these holes; so that, instead of a single current of air frequently created and interrupted, there are a great number of simultaneous currents of air frequently created and interrupted. By observing the character of the note produced, as known to musical ears, and by registering the number of current-interruptions, it is found that corresponding to the note C, which is that of the white key on the left of the two black keys usually next on the right of the lock of a pianoforte, and which note is thus written

the number of current-interruptions is, in modern music, 528 in a second of time. (On variations of this number we shall speak hereafter.)

(3) The Reed. This instrument, in its ruder form, (Figure 12), consists of a small pipe inserted in a larger pipe into which air is driven; the only outlet for the air in the large pipe being through the small pipe; and the only way by which air can enter the small pipe being by a long aperture that is closed by a thin plate or tongue of elastic metal which has a tendency to stand slightly open, leaving a narrow opening opposed to the incoming current of air. As soon as the current is strong, it claps the tongue close; the elasticity of the tongue opens it; it is clapped again, &c. The times of vibration of the tongue are uniform (depending on its elasticity), and a musical note, of rather harsh character, is produced. In the more refined Reeds, the tongue vibrates through an aperture without touching the sides, and then produces a sweeter musical note; this is the construction used in all instruments of the class of the harmonium.

(4) The Monochord, or single stretched string. This has usually been made, for experimental purposes, as a single wire, fixed at one end, passing over two bridges, and stretched by a weight at the other end

When it is wished to keep the wire in a horizontal position, the wire may either be led over a pulley, or may be attached to one arm of a rectangular lever, the other arm carrying the weight. But, if a vertical position is admissible, it is sufficient to suspend a weight freely to it (as the vibration of the weight corresponding to a small vibration of the string is, in practice, quite insensible), care being taken that the wire is tightly nipped at the top and bottom. When such a wire is plucked aside and allowed to vibrate, it gives a musical note: the pitch of the note does not depend on the place of plucking it, but the quality of the tone does depend on it. Upon measuring carefully the length of the wire, the weight of the wire, and the weight which stretches it, the number of vibrations made in a second of time can be computed (the theory of this will be given below). If the extending weight or the length of the string be altered by trial till the string gives a definite note, for instance, the C mentioned above, then it is found that the calculation gives the same number of complete vibrations (a motion backwards and a motion forwards being understood to mean one complete vibration) as the number of passages and interruptions of air in the experiment with the Siren.

All these experiments prove that the formation of a musical note depends on the repetition of similar disturbances of the air at equal intervals of time; but only those of the Siren and the Monochord give the means of computing the frequency of vibrations for an assigned musical note.

72. *The quality of a musical note is determined by the form of the function which expresses the atmospheric disturbance.*

The investigations of Articles 21, &c., 44, &c., 50, &c., have given us expressions for the displacement of the particles of air in the propagation of sound, in all cases represented by $\phi(at-x)$ or by a multiple of $\phi(at-x)$, where the form of the function ϕ is undetermined. In the *Partial Differential Equations*, Article 22, it is shewn that, algebraically, the function ϕ may be in any way discontinuous; and in Article 28, above, are explained the only limitations that physical considerations appear to impose on the generality of discontinuity. We have now another limitation, namely, that the function must be periodical (producing similar disturbances of the air after the repetition of equal intervals of time). With all these limitations, however, it will be seen that there is a very great range in the variations which may be given to the form of the function. But the condition of periodicity gives great facility for the consideration or the determination of the form. Putting v for $at-x$, and supposing that the equal values of $\phi(v)$ return when v is increased by λ, 2λ, &c., it will be seen that such a function may be represented to any degree of approximation by a series of terms, such as

$$A_1 \sin \frac{2\pi}{\lambda} v + A_2 \sin \frac{4\pi}{\lambda} v + \text{\&c.}$$

$$+ B_1 \cos \frac{2\pi}{\lambda} v + B_2 \cos \frac{4\pi}{\lambda} v + \text{\&c.}:$$

which are all periodical when v is increased by a multiple of λ; whose aggregate may amount to 0 or to any assigned value for a given value of v; and in which, if we wish to represent n numerical values of v, we have only to take n terms, and we are able to determine the coefficients, A_1, A_2, B_1, B_2, &c. (See the *Partial Differential Equations*, Article 60, where is explained the process of effecting the determination so as to represent both $\phi(v)$ and $\phi'(v)$). Adopting then this form, it appears that we may have any of the following forms of the function:—

$$A_1 \sin \frac{2\pi}{\lambda} v + B_1 \cos \frac{2\pi}{\lambda} v,$$

which goes through its changes only once while v increases by λ: (This appears to be the function for the tuning-fork.)

$$A_2 \sin \frac{4\pi}{\lambda} v + B_2 \cos \frac{4\pi}{\lambda} v,$$

which goes through its changes twice while v increases by λ:

Similar terms which go through all their changes three times, four times, &c., while v increases by λ:

Any combination of these terms.

Now if we consider the first of these functions as representing the simple form of that disturbance which produces in the ear the sensation of the *fundamental tone*, then the second of these functions will represent vibrations recurring twice as frequently, which we shall find to be very important in music, as representing the *octave above the fundamental tone;* the third will give the *twelfth above the fundamental tone;* the fourth will give the *double octave above the fundamental tone,* &c. These notes have usually been called collectively the *harmonics of the fundamental tone.* (Professor Tyndall has lately advocated the use of the term *overtones,* derived from the German.)

It appears then that every musical note may be represented by a combination of the fundamental tone and its harmonics in some proportion. It seems that only the note produced by the tuning-fork is confined to the first term or fundamental tone*. It seems probable (from consideration of the mechanical movements), that the ancient Reed requires the supposition of a great departure from the simple law of sines, which implies the introduction of large terms of the higher harmonics, greatly impairing the effect of the fundamental tone. In the case of vibrating wires, we know from the mechanical theory (to be given hereafter), that there is always a mixture of harmonics with the fundamental tone; the character of the mixture depending on the point of the wire at which motion is given to it by the finger or the key.

The pure fundamental note, as given by the tuning-fork, is sweet, but somewhat inanimate. Richness is

* It is possible, by an injudicious blow, to produce a different vibration of the tuning-fork, with a very high tone, but it quickly dies away

given by an admixture of the two or three first harmonics.

We shall see hereafter (Article 104), that upon the quality of the note or the form of the function depends the formation of vowel-sounds.

73. *Theory of the vibrations of a musical string.*

We shall take for unit of weight, the weight of length 1 of the vibrating wire. Let the length of the wire between its two points of fixation be l; and let the tension, expressed by a weight referred to the unit above mentioned, be L. We shall suppose the wire to have a small elasticity, so that, in the excessively minute increase of length produced by pulling it aside for vibration, its tension will not sensibly differ from L. For any point of the wire, let x be measured in the straight line from one fixed end towards the other, and let y and z be rectangular ordinates measured from that line; y and z are then the displacements of the point of the wire, produced by the vibration.

For the point whose abscissa is $x-h$, the other coordinates are $y-\frac{dy}{dx}h$, and $z-\frac{dz}{dx}h$ (h being indefinitely small). Hence the two points x and $x-h$ are at the opposite angles of a parallelopiped, whose sides are h, $\frac{dy}{dx}h$, $\frac{dz}{dx}h$, and whose diagonal (in the direction of that portion of the wire) is sensibly equal to h. Therefore the resolved tensions on the point x produced by the antecedent part of the wire are respectively $-L\frac{dy}{dx}$ in the direction y, and $-L\frac{dz}{dx}$ in the direction z. In like manner, using $x+k$ for x, the resolved tensions on the point $x+k$ produced by the following part of the wire are respectively

$$+L\left(\frac{dy}{dx}+\frac{d^2y}{dx^2}\cdot\frac{k}{1}\right) \text{ in the direction } y,$$

and

$$+L\left(\frac{dz}{dx}+\frac{d^2z}{dx^2}\cdot\frac{k}{1}\right) \text{ in the direction } z.$$

The real forces, therefore, tending to move the length k from the axis of x, are

$$+L\frac{d^2y}{dx^2}k \text{ in the direction } y,$$

and

$$+L\frac{d^2z}{dx^2}k \text{ in the direction } z.$$

The mass moved is k.

Hence, putting a^2 for gL*,

$$\frac{d^2y}{dt^2}=gL\frac{d^2y}{dx^2}=a^2\frac{d^2y}{dx^2};$$

$$\frac{d^2z}{dt^2}=gL\frac{d^2z}{dx^2}=a^2\frac{d^2z}{dx^2}.$$

As these two equations are independent and pre-

* In the *Partial Differential Equations*, p. 46, line 3, *read* "where $a^2=gL$." A type has dropped.

cisely similar, it will be sufficient in the first place to examine one only in detail. A portion of the following will be found in the *Partial Differential Equations*, pages 46, &c.

The solution of the second equation is,

$$z=\phi(at-x)+\psi(at+x).$$

For every value of t, this must $=0$ when $x=0$, and also when $x=l$, inasmuch as the string is fixed at these points. Therefore, putting successively for x the values 0 and l,

$$\phi(at)+\psi(at)=0,$$

$$\phi(at-l)+\psi(at+l)=0.$$

We shall call these the terminal equations.

Since the first equation must hold for every value of t, ψ must $=-\phi$; and the second equation is changed to

$$\phi(at-l)-\phi(at+l)=0;$$

and this equation must hold for every value of t, and consequently for every value of $at-l$. Put q for $at-l$; then, whatever be the value of q,

$$\phi(q)-\phi(q+2l)=0.$$

That is to say, the form of ϕ must be such that the value of the function is the same when the quantity under the bracket is increased or diminished by $2l$. In other words, the function must be periodical, going through its changes while the quantity which it affects is changed by $2l$. Or,

$$z=\phi(at-x)-\phi(at+x),$$

where the function ϕ is periodical, going through all its changes and returning to the same value while the quantity in the bracket is altered by $2l$.

In like manner,

$$y = \chi(at - x) - \chi(at + x),$$

where the periodic character of χ is similar to that of ϕ, but where there is no connexion between the forms of the functions ϕ and χ.

In both these expressions, for a given value of x, the functions go through their periodic changes while at is increased by $2l$, or while t is increased by

$$\frac{2l}{a} \text{ or } \frac{2l}{\sqrt{(gL)}}.$$

Hence the time of complete vibration, both in the direction of y and in the direction of z, is $\frac{2l}{\sqrt{(gL)}}$; a calculable quantity, and which has been calculated in experiments made for ascertaining the frequency of vibrations corresponding to a recognized note.

It appears from this investigation that the vibrations of a wire fixed at both ends necessarily recur at equal intervals of time, and therefore necessarily produce a musical note.

The effect of the combination of the values of y and z for any special value of x deserves notice. For all the times τ, $\tau+\frac{2l}{a}$, $\tau+\frac{4l}{a}$, &c., the values of y are the same. For all the same times τ, $\tau+\frac{2l}{a}$, &c., the values of z are the same, though generally different from those of y. Hence, for all the times τ, $\tau+\frac{2l}{a}$, &c., the point of the wire will be in the same position in the plane xy, changing however its position in the intermediate time. Therefore the point of the wire will constantly describe the same orbit in the plane xy. This has been rendered visible by allowing the sun-light to pass through a narrow chink in the plane xy, and to illuminate a point of the wire. The orbit described by the point is sometimes very simple and sometimes fantastically complex. It is thought to be not improbable that these differences depend on the skill of the musician who excites the vibration, and that they produce different qualities of the note.

74. *Nature of the vibrations of a musical string as depending upon its initial circumstances.*

For the moment, put θ for the quantity affected by the functional symbol ϕ. It appears that $\phi(\theta)$ is a periodical function, never infinite in value, never changing its value by a *saltus* with actual disruption of values, and going through all its changes while θ is changed by $2l$. Such a function may always be represented by a sufficient number of integral powers and products of $\sin\frac{\pi\theta}{l}$ and $\cos\frac{\pi\theta}{l}$; and these powers and products may be converted into simple sines and cosines of multiples of θ. Hence the function $\phi(\theta)$ may be represented by

$$\Sigma\left(A_n.\sin\frac{n\pi\theta}{l}\right)+\Sigma\left(B_n.\cos\frac{n\pi\theta}{l}\right);$$

and, putting $at-x$ for θ in order to represent the first term in the expression for z, and $at+x$ for θ in order to represent the second term,

$$z=\Sigma\left\{A_n.\sin\frac{n\pi(at-x)}{l}\right\}$$

$$+\Sigma\left\{B_n.\cos\frac{n\pi(at-x)}{l}\right\}$$

$$-\Sigma\left\{A_n.\sin\frac{n\pi(at+x)}{l}\right\}$$

$$-\Sigma\left\{B_n.\cos\frac{n\pi(at+x)}{l}\right\};$$

or,

$$z=-\Sigma\left(2A_n.\cos\frac{n\pi at}{l}.\sin\frac{n\pi x}{l}\right)$$

$$+\Sigma\left(2B_n.\sin\frac{n\pi at}{l}.\sin\frac{n\pi x}{l}\right),$$

which satisfies the general equations and the terminal equations.

Then the value of $\frac{dz}{dt}$ is

$$+\Sigma\left(\frac{2n\pi a}{l}A_n.\sin\frac{n\pi at}{l}.\sin\frac{n\pi x}{l}\right)$$

$$+\Sigma\left(\frac{2n\pi a}{l}B_n.\cos\frac{n\pi at}{l}.\sin\frac{n\pi x}{l}\right).$$

Put Z and Z' for the values of z and $\frac{dz}{dt}$ when $t=0$; then

$$Z=-\Sigma\left(2A_n.\sin\frac{n\pi x}{l}\right),$$

$$Z'=+\Sigma\left(\frac{2\pi a}{l}nB_n.\sin\frac{n\pi x}{l}\right).$$

Now when Z and Z' are actually given in an algebraical form, the values of A_n and B_n may be found by the following process. Multiply the given expression for Z by $\sin\frac{m\pi x}{l}$, m being any integer, and integrate from $x=0$ to $x=l$; let S_m be the value of the definite integral. Multiply each of the terms $-2A_n.\sin\frac{n\pi x}{l}$ by $\sin\frac{m\pi x}{l}$ for the same purpose. When m differs from n, the product will be

$$-A_n.\cos\frac{(m-n)\pi x}{l}+A_n.\cos\frac{(m+n)\pi x}{l};$$

and the integral of these terms from $x=0$ to $x=l$ is 0. But, when $m=n$, the product is

$$-A_n + A_n \cos\frac{2n\pi x}{l},$$

whose integral from $x=0$ to $x=l$ is $-A_n . l$. Hence $A_n = -\frac{S_n}{l}$.

A similar process applies to the expression for Z'.

It will be interesting to apply this method in some specific cases.

75. *Vibration of a string which has been pulled aside at the center of its length.*

Suppose the central point of a string to be pulled aside through the distance c, and to be then allowed to start from a state of rest. Z' will $=0$ at every point, and therefore the whole series of terms represented by B_n will vanish.

The expression for Z will consist of two parts. From $x=0$ to $x=\frac{l}{2}$, $z=\frac{2cx}{l}$.

From $x=\frac{l}{2}$ to $x=l$, $z=\frac{2cl-2cx}{l}$.

Multiply the first part by $\sin\frac{n\pi x}{l}$, and integrate from $x=0$ to $x=\frac{l}{2}$; the integral is

$$-\frac{cl}{n\pi}\cos\frac{n\pi}{2}+\frac{2cl}{n^2\pi^2}\sin\frac{n\pi}{2},$$

Multiply the second part by $\sin\frac{n\pi x}{l}$, and integrate from $x=\frac{l}{2}$ to $x=l$; the integral is

$$+\frac{cl}{n\pi}\cos\frac{n\pi}{2}+\frac{2cl}{n^2\pi^2}\sin\frac{n\pi}{2}-\frac{2cl}{n^2\pi^2}\sin n\pi.$$

The sum gives

$$S_n=\frac{cl}{\pi^2}\left(\frac{4}{n^2}\sin\frac{n\pi}{2}-\frac{2}{n^2}\sin n\pi\right).$$

And, forming the values of $-A_n$ or $\frac{S_n}{l}$ for each of the values $n=1$, $n=2$, &c., and then completing the formula, we obtain,

$$A_1=\frac{-4c}{\pi^2},\quad A_2=0,\quad A_3=\frac{+4c}{9\pi^2},$$

$$A_4=0,\quad A_5=\frac{-4c}{25\pi^2},\ \text{\&c.};$$

and $Z=\frac{8c}{\pi^2}\left(\sin\frac{\pi x}{l}-\frac{1}{9}\sin\frac{3\pi x}{l}+\frac{1}{25}\sin\frac{5\pi x}{l}-\text{\&c.}\right)$,

which is the equation for the two sides of the triangle, in the form suiting our purpose.

(The attention of the algebraical student is invited to the circumstance, that the equation to two discontinuous straight lines of limited extent is given by one algebraical expression for the value of the ordinate at every point*. The same remark applies in the next example.)

Introducing the values of A_1, A_2, &c., into the general expression, we have for z,

$$\frac{8c}{\pi^2}\left(\sin\frac{\pi x}{l}\cos\frac{\pi at}{l}-\frac{1}{9}\sin\frac{3\pi x}{l}\cos\frac{3\pi at}{l}+\frac{1}{25}\sin\frac{5\pi x}{l}\cos\frac{5\pi at}{l}-\&c.\right).$$

The principal note (that is, the vibration with the largest coefficient) is given by $\cos\frac{\pi at}{l}$, or is the funda-

* To verify this theorem numerically, we have supposed the base of the triangle divided into 36 equal parts, and have calculated at every dividing point the value of the ordinate from the formula above, using eight terms of the series; and have compared it with the triangular ordinate. The comparison is as follows (omitting the factor c, and multiplying by 100):

Triangular ordinate,
0, 6, 11, 17, 22, 28, 33, 39, 44, 50, 56, 61, 67, 72, 78, 83, 89, 94,
Ordinate by formula,
0, 5, 11, 17, 22, 28, 33, 39, 45, 50, 55, 61, 67, 72, 77, 83, 89, 95,

(comparison continued),
100, 94, 89, 83, 78, 72, 67, 61, 56, 50, 44, 39, 33, 28, 22, 17, 11, 6, 0,
98, 95, 89, 83, 77, 72, 67, 61, 55, 50, 45, 39, 33, 28, 22, 17, 11, 5, 0.

The student will remark in the formula of the text that at the middle of the string, where $\frac{x}{l}=\frac{1}{2}$, every trigonometrical term $=+1$. Thus we have an incidental proof of the theorem,

$$1=\frac{8}{\pi^2}\left\{1+\frac{1}{9}+\frac{1}{25}+\&c.\right\}.$$

mental note. There are none of the even harmonics, depending on $\cos\frac{2\pi at}{l}$, $\cos\frac{4\pi at}{l}$, &c., but there are all the odd harmonics.

76. *Vibration of a string which has been pulled aside at an excentric point.*

Suppose, as an instance, the string to be plucked aside at the distance $\frac{3l}{4}$ from the first point, and to be allowed to start from a state of rest.

As in the last example, the train of terms represented by B_n will vanish. To find the values of A_n we remark that; from $x=0$ to $x=\frac{3l}{4}$, $z=\frac{4cx}{3l}$; and from $x=\frac{3l}{4}$ to $x=l$, $z=c\times\left(4-\frac{4x}{l}\right)$.

Multiplying the first of these values by $\sin\frac{n\pi x}{l}$, and integrating from $x=0$ to $x=\frac{3l}{4}$, the integral is

$$-\frac{cl}{n\pi}\cos\frac{3n\pi}{4}+\frac{4cl}{3n^2\pi^2}\sin\frac{3n\pi}{4}.$$

Multiplying the second value by $\sin\frac{n\pi x}{l}$, and integrating from $x=\frac{3l}{4}$ to $x=l$, the integral is

$$+\frac{cl}{n\pi}\cos\frac{3n\pi}{4}+\frac{4cl}{n^2\pi^2}\sin\frac{3n\pi}{4}-\frac{4cl}{n^2\pi^2}\sin n\pi.$$

The whole integral, or the sum of the two parts, is

$$S_n = \frac{16cl}{3n^2\pi^2}\sin\frac{3n\pi}{4} - \frac{4cl}{n^2\pi^2}\sin n\pi\,;$$

and therefore

$$-2A_n \text{ or } \frac{2S_n}{l} = \frac{32c}{3n^2\pi^2}\sin\frac{3n\pi}{4} - \frac{8c}{n^2\pi^2}\sin n\pi$$

$$= \frac{32c}{3\pi^2}\left(\frac{1}{n^2}\sin\frac{3n\pi}{4} - \frac{3}{4n^2}\sin n\pi\right).$$

Giving to n the successive values 1, 2, &c., the value of z is found to be

$$\frac{32c}{3\pi^2}\left\{\sqrt{\frac{1}{2}}\sin\frac{\pi x}{l}\cos\frac{\pi at}{l} - \frac{1}{4}\sin\frac{2\pi x}{l}\cos\frac{2\pi at}{l}\right.$$

$$+\sqrt{\frac{1}{2}}\cdot\frac{1}{9}\sin\frac{3\pi x}{l}\cos\frac{3\pi at}{l}$$

$$-\sqrt{\frac{1}{2}}\cdot\frac{1}{25}\sin\frac{5\pi x}{l}\cos\frac{5\pi at}{l} + \frac{1}{36}\sin\frac{6\pi x}{l}\cos\frac{6\pi at}{l}$$

$$\left.-\sqrt{\frac{1}{2}}\cdot\frac{1}{49}\sin\frac{7\pi x}{l}\cos\frac{7\pi at}{l} + \&c.\right\},$$

in which series every 4th term vanishes. When $t=0$, it gives for Z an expression possessing the same peculiarity which we have remarked in the first example.

In this general value for z it will be seen that, while the vibration of the largest coefficient is that of the fundamental note $\left(\text{depending on } \cos\frac{\pi at}{l}\right)$, yet the first even harmonic $\left(\text{depending on } \cos \frac{2\pi at}{l}\right)$ has a considerable coefficient; in which respect the instance before us differs remarkably from the first example. It is the practice, we believe, of all violinists to touch the strings of the violin with the bow, not at the middle (as in the first example) but at nearly three-quarter-length (as in the second example), or even much nearer to the end of the string. In selection of the point at which each key-lever of the piano-forte strikes its wire, we believe that the best makers adopt a similar principle. The reason appears to be that, as we have remarked at the end of Article 72, a richness which is pleasurable (we know not why) is produced by the combination of some of the harmonics with the fundamental note; and that this condition is far better secured, as appears in the investigations which we have just made, by the excentric disturbance of the string than by the central disturbance.

In the harp and various other instruments, the string is once plucked aside to produce a note; in the piano-forte it is once struck; in the violin it is, apparently, pulled aside several times in a second by the light touch of the bow, armed with a substance that produces sufficient friction. All these cases are equally comprehended in the theory above.

77. *Importance of the connexion of musical strings with a sounding-board.*

It has been seen in Article 48 that in a divergent oscillating wave of air, such as we may suppose to be caused by the vibrations of a string, the motion of the particles is of the order of *R*, whose first term varies as the distance raised to the power $-\frac{1}{2}$. Moreover, the smallness of dimension of a wire makes it impossible that it can communicate great motion even to the air which it touches. Hence, it is impossible that a wire can, by immediate action on the air, produce a sound easily audible to a considerable or convenient distance. To make it audible, the wire must be connected with an intermediate substance whose vibrations can produce a stronger effect on the air, and those vibrations must be excited by the vibrations of the wire. The intermediate substance used for this purpose is the sounding-board.

In the violin, the wires pass over a bridge which rests by two feet upon the upper board; and under that board, at the place where one foot of the bridge presses, is a little post (known by the name of the "sound-post" or the "soul") connecting the upper board with the lower board. Every tremulous motion of a wire of the violin acts directly upon the bridge and upon the upper and lower boards; and the tremors of these produce effective vibrations of the air, and diffuse the sound. We know that every thing depends on the elastic properties of these boards; but we know nothing of their precise laws of vibration.

In the piano-forte, the general construction is simpler, but the sounding board is so connected with the supports of the wires that it is made to vibrate by the vibrations of the wires.

Though we cannot give a theory which shall apply accurately to the motions of the sounding-board, we can give one founded on motions which have a certain degree of analogy with the motions of the wire and sounding-board. Consider the wire as a pendulum, whose length is L and weight W, and the sounding-board also as a pendulum, whose length is l and weight w; and suppose these two pendulums to be connected. When both are displaced through the same space z, the pressure-force tending to bring them back is

$$W\frac{z}{L}+w\frac{z}{l};$$

the mass to be moved is $W+w$; the equation of motion is therefore

$$\frac{d^2z}{dt^2}=-g\,\frac{\frac{W}{L}+\frac{w}{l}}{W+w}\,z;$$

its solution is

$$z=A\sin(ct+B),$$

where $$c^2 = g\frac{\frac{W}{L}+\frac{w}{l}}{W+w};$$

and the time of a complete vibration, in which $\sin(ct+B)$ goes through all its changes, is

$$\frac{2\pi}{c} = 2\pi\sqrt{\frac{Ll\,(W+w)}{g\,(Wl+wL)}}.$$

The time of complete vibration of the first pendulum, if not connected with the second, would be $2\pi\sqrt{\frac{L}{g}}$. Hence the time of vibration is altered, by the connection, in the proportion of 1 to $\sqrt{\frac{Wl+wl}{Wl+wL}}$. If T be the time of complete vibration of the first pendulum, and τ that of the second, supposed to be unconnected, the time of vibration of the first pendulum is altered by the connection in the proportion of 1 to $\sqrt{\frac{W\tau^2+w\tau^2}{W\tau^2+wT^2}}$. If, in one combination, the first pendulum vibrates in T, the connection-alteration is as 1 to $\sqrt{\frac{4W\tau^2+4w\tau^2}{4W\tau^2+4wT^2}}$; if, in another combination, it vibrates in $\frac{T}{2}$, the connection-alteration is as 1 to $\sqrt{\frac{4W\tau^2+4w\tau^2}{4W\tau^2+wT^2}}$. Hence it appears possible that the frequency of vibration may be altered in different proportions for the fundamental note and for its harmonics; and the series of sounds which reach the ear may not possess the character of harmonics. The pleasurable character of the complex sound will therefore depend entirely upon laws of vibration of the sounding-board, which we are unable to investigate mathematically, but which perhaps in some cases are rudely mastered in practice.

We may add to this subject that, in the common tuning-fork, the separation and approximation of the two branches appear to produce a longitudinal retraction and extrusion of the central fibres of the stalk; and, when the stalk is planted downwards upon a table, the sound of the fork is very much increased; the table acting as a sounding-board. In this manner it is commonly used by the tuners of musical instruments.

78. *Theory of the vibrations of air in an organ-pipe stopped at both ends.*

In Article 23 we found as the general expression for the disturbance of air in a tube (omitting, for convenience, the factors n and θ),

$$X = \phi(at - x) + \psi(at + x);$$

from which

$$\frac{dX}{dt} = a \cdot \phi'(at - x) + a \cdot \psi'(at + x).$$

The further treatment of these formulæ requires us to distinguish three cases: (1), that of a tube stopped at both ends; (2), that of a tube open at both ends;

(3), that of a tube stopped at one end and open at the other.

When a tube is stopped at both ends (it being supposed that adequate means are provided for putting the air into a state of vibration) the condition required is, that the disturbance at both ends is 0. Let the length be l, and let x be measured from one end; then, exactly as in Article 73,

$$\phi(at) + \psi(at) = 0,$$

$$\phi(at - l) + \psi(at + l) = 0;$$

the terminal equations in this case; from which it appears that $\psi = -\phi$, and that ϕ is a periodic function going through all its changes while at increases by $2l$, or while t increases by $\frac{2l}{a}$. The number of complete vibrations per second will be $\frac{a}{2l}$. Now on comparing the expression in Article 23 with the discussion in Article 24, it appears that $n\theta a$, for which (with a convenient abbreviation) we have here used a, is the space described by external sound in one second of time. Hence the number of complete vibrations per second of the air in the closed organ-pipe is

$$\frac{\text{velocity of sound in feet per second}}{2 \times \text{length of pipe in feet}}.$$

In ordinary temperatures the velocity of sound is about 1090 feet per second (Article 65); a pipe 1·03 foot long will give 528 vibrations per second; a number to which we shall hereafter refer.

The periodic function ϕ, just as in Article 74, may contain terms depending on

$$\begin{matrix}\cos\\ \sin\end{matrix}\frac{\pi at}{l},\quad \begin{matrix}\cos\\ \sin\end{matrix}\frac{2\pi at}{l}\ \&c.,$$

satisfying the original equations and the terminal equations, and giving a fundamental note and its harmonics; and, with proper exciting causes, any one of these may exist, to the exclusion of the others. For instance, if we have a pipe 2·06 feet long, and if we excite the air by a disturbance recurring 528 times in a second, it will not produce waves represented by $\frac{\pi at}{l}$ recurring 264 times in a second (which gives the fundamental note for that pipe), but will produce waves represented by $\frac{2\pi at}{l}$ recurring 528 times in a second (giving the first harmonic for that pipe). And so for higher harmonics.

The velocity of a particle, as in Article 74, is expressed by

$$\Sigma\left(\frac{2n\pi a}{l}A_n.\sin\frac{n\pi at}{l}.\sin\frac{n\pi x}{l}\right)$$

$$+\Sigma\left(\frac{2n\pi a}{l}B_n.\cos\frac{n\pi at}{l}.\sin\frac{n\pi x}{l}\right).$$

Now, when the fundamental note alone is sounded, or when the only terms employed are those depending on

$\frac{\pi at}{l}$, that is when $n = 1$, those terms are necessarily associated with the factor $\sin \frac{\pi x}{l}$; and from $x = 0$ to $x = l$ there is no point of the tube where this factor vanishes. Therefore, in no part of the tube is the air at rest. But, if the first harmonic alone is sounded, or when the only terms employed are those depending on $\frac{2\pi at}{l}$, that is when $n = 2$, those terms are necesarily associated with the factor $\sin \frac{2\pi x}{l}$; and this factor vanishes when $x = \frac{l}{2}$, or at the middle of the pipe's length. There is therefore a point of absolute rest at the middle of the length: this is called a *node*. In like manner, if the second harmonic alone is sounded, there are two *nodes*, dividing the length into three equal parts; and so for higher harmonics.

79. *Theory for a pipe open at both ends.*

When a pipe is open at both ends, the algebraical conditions are totally different. The physical condition to be satisfied is that, either at the pipe's mouth, or more probably at a small distance exterior to it, the pressure and density of the air are the same as the pressure and density of the tranquil atmosphere. Now in Article 21 it is found that the variable part of Δ is $-D\frac{dX}{dx}$, or

$$D . \phi'(at - x) - D . \psi'(at + x).$$

Making this $=0$ when $x=0$ and when $x=l$,

$$\phi'(at) - \psi'(at) = 0$$

$$\phi'(at - l) - \psi'(at + l) = 0,$$

which are the terminal equations for this case.

The first equation, which is general for all values of t, gives $\psi' = \phi'$. The second gives

$$\phi'(at - l) = \psi'(at + l),$$

which, as in Article 73, shews that ϕ' is a periodical function, going through all its changes while the quantity affected by it is changed by $2l$, or while t increases by $\frac{2l}{a}$: giving the same number of complete vibrations per second, and therefore the same fundamental note, as a pipe closed at both ends, Article 78. The velocity of the particle, or

$$a \,.\, \phi'(at - x) + a \,.\, \psi'(at + x),$$

may be represented (for the same reasons as in the beginning of Article 74) by

$$\Sigma \left\{ C_n \,.\, \sin \frac{n\pi (at - x)}{l} \right\}$$

$$+ \Sigma \left\{ D_n \,.\, \cos \frac{n\pi (at - x)}{l} \right\}$$

$$+ \Sigma \left\{ C_n \,.\, \sin \frac{n\pi (at + x)}{l} \right\}$$

$$+ \Sigma \left\{ D_n \,.\, \cos \frac{n\pi (at + x)}{l} \right\};$$

or
$$\Sigma\left(2C_n.\sin\frac{n\pi at}{l}.\cos\frac{n\pi x}{l}\right)$$
$$+\Sigma\left(2D_n.\cos\frac{n\pi at}{l}.\cos\frac{n\pi x}{l}\right).$$

The actual displacement of the particle will be found by integrating with respect to t.

Here, when the fundamental note alone is sounded, that is, when $\frac{n\pi at}{l}$ has the smallest factor of t, or $n = 1$, this term is connected with the factor $\cos\frac{\pi x}{l}$, and that factor vanishes when $\frac{\pi x}{l} = \frac{\pi}{2}$, or $x = \frac{l}{2}$. That is, there is perpetual quiescence of air, or a *node*, at the middle of the pipe's length.

In like manner, when the first harmonic is sounded, there will be nodes at the distances $\frac{l}{4}$ and $\frac{3l}{4}$; when the second harmonic is sounded, there will be nodes at the distances $\frac{l}{6}$, $\frac{3l}{6}$, $\frac{5l}{6}$; and so on.

•80. *Theory for a pipe closed at one end and open at the other.*

When a pipe is closed at one end and open at the other, the conditions are more complicated. Suppose the closed end to be that where $x=0$; then, taking the general expression for X as in Article 23,

$$X=\phi(at-x)+\psi(at+x),$$

and consequently the variable part of Δ

$$=D.\phi'(at-x)-D.\psi'(at+x),$$

we must have, when $x=0$, $X=0$, or

$$\phi(at)+\psi(at)=0;$$

when $x=l$, variable part of $\Delta=0$, or

$$\phi'(at-l)-\psi'(at+l)=0.$$

The first equation being perfectly general with respect to t, we may differentiate it, and we find

$$a\phi'(at)+a\psi'(at)=0.$$

This equation, with that immediately preceding, are the terminal equations here.

From the last, $\psi'=-\phi'$; and making this change of function in the second equation,

$$\phi'(at-l)+\phi'(at+l)=0;$$

or, putting q for $at-l$,

$$\phi'(q)+\phi'(q+2l)=0.$$

This equation is different in form from that at which we arrived in Articles 73, 78, and 79; and it shews that, generally, the function ϕ' does *not* go through periodic changes, while the quantity under the function is altered by any constant amount. But a simple treatment gives us the real purport of the equation. Since q may have any value, we may put r in place of q, thus obtaining the equation

$$\phi'(r) + \phi'(r + 2l) = 0;$$

and we may give to r the value $q + 2l$, which converts the equation into

$$\phi'(q + 2l) + \phi'(q + 4l) = 0.$$

Thus we have

$$\begin{aligned} \phi'(q) + \phi'(q + 2l) \qquad\qquad\qquad &= 0, \\ - \phi'(q + 2l) - \phi'(q + 4l) &= 0; \end{aligned}$$

adding the two equations,

$$\phi'(q) - \phi'(q + 4l) = 0;$$

from which we infer that ϕ' is a periodical function which goes through all its changes while the quantity under the function is increased by $4l$. The function of sines and cosines satisfying this condition must be a function of $\frac{\sin}{\cos} \frac{2\pi q}{4l}$, or of $\frac{\sin}{\cos} \frac{\pi q}{2l}$.

Hence $-\phi'(at - x)$, by the reasoning of Article 74, will be expressible by

$$-\Sigma\left\{C_n . \sin\frac{n\pi(at - x)}{2l}\right\} - \Sigma\left\{D_n . \cos\frac{n\pi(at - x)}{2l}\right\}.$$

Then $\qquad +\psi'(at + x) = -\phi'(at + x)$

$$= -\Sigma\left\{C_n . \sin\frac{n\pi(at + x)}{2l}\right\} - \Sigma\left\{D_n . \cos\frac{n\pi(at + x)}{2l}\right\},$$

and $-\phi'(at-x)+\psi'(at+x)$

$$=-\Sigma\left(2C_n.\sin\frac{n\pi at}{2l}.\cos\frac{n\pi x}{2l}\right)$$
$$-\Sigma\left(2D_n.\cos\frac{n\pi at}{2l}.\cos\frac{n\pi x}{2l}\right).$$

Making $x=l$, or $\frac{n\pi x}{2l}=\frac{n\pi}{2}$, we find that the first terminal equation,

$$-\phi'(at-l)+\psi'(at+l)=0,$$

can only be satisfied by making $\cos\frac{n\pi}{2}=0$, or $\frac{n\pi}{2}$ an odd multiple of $\frac{\pi}{2}$; or the values of n must be confined to the odd series 1, 3, 5, &c.

The value of $\phi'(at)+\psi'(at)$, or $\phi'(at)-\phi'(at)$, necessarily satisfies the second terminal equation.

The value of $\frac{dX}{dt}$, or $a.\phi'(at-x)+a.\psi'(at+x)$, is now

$$-a.\Sigma\left(2C_n.\cos\frac{n\pi at}{2l}.\sin\frac{n\pi x}{2l}\right)$$
$$+a.\Sigma\left(2D_n.\sin\frac{n\pi at}{2l}.\sin\frac{n\pi x}{2l}\right);$$

and, integrating with respect to t,

$$X=-\Sigma\left(C_n.\frac{4l}{n\pi}.\sin\frac{n\pi at}{2l}.\sin\frac{n\pi x}{2l}\right)$$
$$-\Sigma\left(D_n.\frac{4l}{n\pi}.\cos\frac{n\pi at}{2l}.\sin\frac{n\pi x}{2l}\right);$$

where n may be any of the odd numbers 1, 3, 5, &c.

For the fundamental note, make $n=1$, and the expression for X goes through all its changes while $\frac{\pi at}{2l}$ changes through 2π, or while t is changed by $\frac{4l}{a}$. We have found in the investigations for a pipe closed at both ends or open at both ends, that the changes occur while t is changed by $\frac{2l}{a}$. Hence, with one end open and one closed, the vibrations of air per second are only half in number of those produced by a pipe of the same length with both ends open or both ends closed. (We shall express this hereafter by saying that its tone is an Octave lower.) Or we may state the theorem thus; the number of vibrations per second in a pipe with one end open and one closed is the same as in a pipe of double the length with both ends open or both ends closed.

The pipe under consideration can give none of the even harmonics, but can give any of the odd ones.

Nodes or quiescent points can only occur when $X=0$, or in this instance, when $\sin\frac{n\pi x}{2l}=0$. (It will be remembered that x is measured from the closed end.) For the fundamental note, $\frac{\pi x}{2l}$ must $=\pi$, or $x=2l$, which is impossible, and there is therefore no node. For the harmonic with $n=3$, $\frac{3\pi x}{2l}$ must $=\pi$, which makes x at a node $=\frac{2l}{3}$. For the harmonic with $n=5$, $\frac{5\pi x}{2l}$ must

$=\pi$ or 2π, which makes x at a node $=\frac{2l}{5}$, or $=\frac{4l}{5}$; and so for higher harmonics.

In instruments of the class of the flute and clarionet, there are holes at definite points, which establish such a communication with the external air, that the variations of elastic pressure in the corresponding parts of the internal column are very small and may be neglected. The student, acting on this suggestion, will be able to adapt the formulæ which we have introduced above, to the case of these instruments.

Wertheim, in the *Comptes Rendus*, tome XXXIII, and the *Annales de Chimie*, 3[d] series, tome XXXI, has given formulæ (undemonstrated) for the effect of the transversal dimensions of organ-pipes.

81. *Methods of exciting the vibrations of air in a musical pipe; theory of Resonance.*

In some pipes, as in the reed-tubes of an organ, in the clarionet, &c., the sounds are produced by forcing a comparatively small stream of air through a subsidiary instrument (the reed) which itself produces musical vibrations; in other pipes, as in organ-pipes with the ordinary mouth-piece, in the flute, &c., the sound is produced by a blast of air which appears to possess nothing whatever musical in its character. To illustrate the effects of these, and more particularly the effect of the reed, the following case will be investigated. Suppose that we have a pipe closed at one end, and suppose that by an external action the air at the other end is compelled to move according to the law $b \,.\, \sin ct$; to find the motion of the air in the pipe.

The formula for the motion of the air, given by the solution of the differential equations, is

$$X = \phi(at - x) + \psi(at + x),$$

where we shall measure x from the closed end of the pipe. The terminal conditions are,

When $x = 0$, $X = 0$, or $\phi(at) + \psi(at) = 0$,
for all values of t.

When $x = l$, $X = b \,.\, \sin ct$,

$$\text{or } \phi(at - l) + \psi(at + l) = b \,.\, \sin ct.$$

The first equation gives $\psi = -\phi$; and the second then becomes

$$\phi(at - l) - \phi(at + l) = b \,.\, \sin ct.$$

It appears necessary, in the first place, that the multiple of t under the function be ct; which will be effected by putting the equation under this form:

$$\chi\left\{\frac{c}{a}(at - l)\right\} - \chi\left\{\frac{c}{a}(at + l)\right\} = b \,.\, \sin ct,$$

$$\text{or } \chi\left(ct - \frac{cl}{a}\right) - \chi\left(ct + \frac{cl}{a}\right) = b \,.\, \sin ct.$$

In the next place, it appears impossible that any function except a sine or cosine on the left hand can produce the sine on the right hand. Suppose then

$$\chi = h \times \text{sine} + k \times \text{cosine}.$$

Then

$$\left.\begin{array}{r} h.\sin\left(ct-\frac{cl}{a}\right)+k.\cos\left(ct-\frac{cl}{a}\right) \\ -h.\sin\left(ct+\frac{cl}{a}\right)-k.\cos\left(ct+\frac{cl}{a}\right) \end{array}\right\} = b.\sin ct.$$

Or $\quad -2h.\cos ct.\sin\frac{cl}{a}+2k.\sin ct.\sin\frac{cl}{a}=b.\sin ct.$

Hence $h=0$ (unless $\sin\frac{cl}{a}=0$), and $k=\dfrac{b}{2\sin\frac{cl}{a}}$.

Therefore (as $\chi = h \times \text{sine} + k \times \text{cosine} = k \times \text{cosine}$),

$$X=\chi\left(ct-\frac{cx}{a}\right)-\chi\left(ct+\frac{cx}{a}\right)$$

$$=\frac{b}{2\sin\frac{cl}{a}}\left\{\cos\left(ct-\frac{cx}{a}\right)-\cos\left(ct+\frac{cx}{a}\right)\right\}.$$

Or, $\qquad X=\dfrac{b}{\sin\frac{cl}{a}}.\sin ct.\sin\dfrac{cx}{a}.$

(It is possible that, with this term, there may be combined other terms of the nature of those found in Article 78; but they do not affect the present investigation.)

From this expression it appears,

First, that the air in the pipe is made to vibrate in the time corresponding to the time of vibration of the external cause (reed, &c.), even though that time do not agree with the natural time of vibration in the pipe, as found by preceding investigations. And therefore the note given by the pipe is not its natural note, but the note of the external cause.

Second, that if $\frac{cl}{a}$ be less than π, (in which case $\sin\frac{cx}{a}$ never changes sign,) but not much less than π, then the coefficient of vibration of air, namely,

$$\frac{b \cdot \sin\frac{cx}{a}}{\sin\frac{cl}{a}},$$

becomes for all the middle parts of the pipe extremely large.

Third, that the condition that $\frac{cl}{a}$ be very little less than π, or c very little less than $\frac{\pi a}{l}$, implies that the time of complete vibration of the external cause, which is $\frac{2\pi}{c}$, must be very little greater than $\frac{2l}{a}$, or very little greater than that of the natural vibration in the pipe as closed at both ends or open at both ends.

Hence, if a reed, &c. be blown so as to produce a note a very little below that of an organ-pipe (as closed at both ends or open at both ends), and be attached to one end of that pipe, the other end of the pipe being closed, the reed will produce a very loud sound, of its own pitch.

This is the theory of Resonance. It applies to rooms, &c. which are capable of returning a distinct note, as well as to pipes.

The student will have no difficulty in applying the same principles to other cases. Thus, if the mouth of the pipe be open (as usually happens), the variable part of the expression for density must be made to vanish, for all values of t, when $x=0$. The value thus found for X is

$$\frac{b}{\cos\frac{cl}{a}} \cdot \sin ct \cdot \cos\frac{cx}{a}.$$

(This may be accompanied with terms similar to those found in Article 80.)

82. *Reaction of the air in an organ-pipe upon the reed, &c.*

We shall now investigate, for the case of the closed pipe, the pressure which the air vibrating in the tube impresses on the reed or other agent that acts to put it in motion.

In Article 21, putting Q for $H'DN$, the variable

part of the elasticity of the air is found to be $-Q.\frac{dX}{dx}$. Putting for X the value

$$\frac{b}{\sin\frac{cl}{a}}.\sin\frac{cx}{a}.\sin ct,$$

the variable part of the elasticity is

$$-\frac{cbQ}{a.\sin\frac{cl}{a}}.\cos\frac{cx}{a}.\sin ct;$$

and, when $x=l$, the variable part of the elasticity is

$$-\frac{cbQ}{a}.\text{cotan}\frac{cl}{a}.\sin ct.$$

The vibrating mechanism (the tongue of the reed, for instance) is necessarily beyond the air of the organ-pipe, as measured in the direction of x; and the external atmospheric air produces no sensibly varying pressure on its external surface. The force on the reed-tongue, produced by the elasticity of the air of the organ-pipe, is therefore

$$-\frac{cbQ}{a}.\text{cotan}\frac{cl}{a}.\sin ct;$$

which, since $\text{cotan}\frac{cl}{a}$ is negative (as $\frac{cl}{a}$ is little less than π), may be represented by $+C.\sin ct$. The elastic force of the reed-tongue itself, which produces the motion $b.\sin ct$, is generally $-c^2\times$ displacement of the tongue, or (with this vibration, whose amplitude is b) is $-c^2b \,.\, \sin ct$. Consequently, the effect of the reaction of the organ-pipe-air upon the reed-tongue is, to reduce the magnitude of the elastic force to $-(c^2b - C) \,.\, \sin ct$, or to diminish the vibrating force upon the reed-tongue. And if it is supposed, as a fundamental condition, that the vibrations of the reed-tongue are always to be made in the one certain time $\frac{2\pi}{c}$, a constant power must be exerted upon the reed, to enable it to keep up its vibrations, much greater than is required when it is not connected with the organ-pipe.

But if (as appears to be the case in practice) the vibrations of the reed-tongue are determined by the combination of its own elastic force with the reaction of the organ-pipe-air, the results are greatly changed. Assume that the law of vibration may be changed from $\sin ct$ to $\sin c't$. The effect of combining $+ C \,.\, \sin c't$ with $-c^2b \,.\, \sin c't$, is to make the entire force acting on the reed-tongue

$$= -(c^2b - C) \sin c't,$$

or to make the force that tends to bring it to its quiescent point weaker than before; and this will make its vibrations slower. Suppose on the one hand the vibrations are so slow that c' is not much greater than $\frac{\pi a}{2l}$, or $\frac{c'l}{a}$ is not much greater than $\frac{\pi}{2}$. Then the reaction force on the reed-tongue, or

$$-\frac{cbQ}{a}\cot\text{an}\,\frac{cl}{a}\,.\sin c't,$$

has a small positive factor, or the entire force will still be of the character $-(c^2b - \text{a small quantity}) \times \sin c't$, and the vibrations will still be made a little slower. Suppose, on the other hand, they are so slow that c' is not much less than $\frac{\pi a}{2l}$, or $\frac{c'l}{a}$ is not much less than $\frac{\pi}{2}$. Then the entire force on the reed-tongue will be of the character $-(c^2b + \text{a small quantity}) \times \sin c't$, and the vibrations will be made a little quicker. The consequence of these actions will be, that the vibrations of the reed-tongue will be made such that the factor of t will become almost equal to $\frac{\pi a}{2l}$, or that its time of complete vibration $\left(\text{namely } \frac{2\pi}{c'}\right)$, and the time of complete vibration of the air in the pipe, will be almost equal to $\frac{4l}{a}$; which, as we have seen, Article 73, is exactly the same as the time of vibration in the pipe without any attached reed. This, it will be remarked, is essentially founded on the supposition that the reed-tongue is so flexible as to permit the reaction of the organ-pipe-air in some measure to control its vibrations.

The symbolical investigation will be as follows. If the reed-tongue were isolated, its vibration would be represented by $b \,.\sin ct$; which indicates that the reed-force is, in all cases, $-c^2 \times$ the ordinate. Hence, if its actual vibration is $b \,.\, \sin c't$, the reed-force is $-bc^2 \,.\, \sin c't$, and the entire force which acts on it is

$$-bc^2 \,.\, \sin c't - \frac{c'bQ}{a} \,.\, \text{cotan}\,\frac{c'l}{a} \,.\, \sin c't.$$

Hence we must have, on the mechanical principles of vibration,

$$-bc^2 \,.\, \sin c't - \frac{c'bQ}{a} \,.\, \text{cotan}\,\frac{c'l}{a} \,.\, \sin c't = -bc'^2 \,.\, \sin c't.$$

$$\text{Or } c'^2 - c^2 = \frac{c'Q}{a} \,.\, \text{cotan}\,\frac{c'l}{a} = \frac{Q}{l} \,.\, \frac{c'l}{a} \,.\, \text{cotan}\,\frac{c'l}{a}.$$

Now, in the supposed case that neither $\frac{cl}{a}$ nor $\frac{c'l}{a}$ differs much from $\frac{\pi}{2}$, let

$$\frac{cl}{a} = \frac{\pi}{2} + e, \quad \frac{c'l}{a} = \frac{\pi}{2} + x;$$

then $c'^2 - c^2$, to the first order of small quantities, is $\frac{\pi a^2}{l^2}(x - e)$, and cotan $\frac{c'l}{a} = -\tan x = -x$, and $\frac{c'l}{a} = \frac{\pi}{2}$; and the equation becomes

$$\frac{\pi a^2}{l^2}(x - e) = -\frac{Q}{l} \cdot \frac{\pi}{2} \,.\, x;$$

from which x is determined, and, by substitution, c' is found.

The tone given by the pipe does therefore, even in very favourable cases, depend in some degree upon the stiffness of the reed.

When the time of vibration is nearly equal to $\frac{4l}{a}$ (the natural time of vibration in the pipe), the coefficient of vibration is nowhere sensibly greater than at the point where the reed acts on it.

There is reason to think that some addition is yet required to this theory. The results at which we have arrived imply that a simple vibration is produced in the air, whose quality, notwithstanding the non-coincidence of the times of independent vibration of the reed and the organ-pipe, is always the same. But the experiments on vowel-sound, to be mentioned below (Article 104), prove that it is not always the same. Possibly, when the mathematical calculus is farther advanced, this may be shewn to depend on the circumstance, that the reed does not occupy the whole breadth of the tube, and waves of different period may be passing at the same time.

83. *Production of musical sound by a simple blast of air.*

The investigation of the reaction on the reed seems to throw some light upon that obscure subject, the production of musical vibrations in a pipe by a simple blast of air. In the ordinary mouth-piece of an organ-pipe, of which a section is given in Figure 13, a strong blast is forced through a very narrow slit, and is received upon a sharp edge, after which it partly enters the body of the pipe, partly passes into the external air. In the flute, a blast of air is directed by the lips of the flute-player upon the sharp edges of a hole in the tube, and then partly enters the tube. In both cases it appears necessary to suppose that the air, which enters the tube, bears a vibration or many kinds of simultaneous vibrations (which as mixed could not be distinguished from ordinary noise). The same supposition appears necessary to explain the sounding of a stretched wire opposite a chink of a door, or the singing of telegraph-wires, or the whistle of a locomotive (in which an annular jet of stream is thrown upon the circular edge of a bell, and excites the note peculiar to the bell). Every one of these vibrations may be considered as the vibration of a reed-tongue; and the reaction of the air in the pipe will modify these in the way which we have described for the reed. There appears to be only this possible difference; that these external air-vibrations have not that stubborn attachment to arbitrary times of vibration which the reed-tongue has, and therefore every one of them will be so changed as to correspond exactly to the vibration natural to the organ-pipe.

A skilful flute-player, making no alteration in the fingering of the holes, but altering the character of his blast, can produce not only the first note but any one of several of its harmonics. Here it appears to be necessary that the external vibrations should have an approximate similarity to those in the note which is to be produced: since the same mode of blowing which produces one note will not produce another.

The matter however demands more complete explanation.

It is scarcely necessary to say that the energy of the vibrations of the air in the organ-pipes is consumed in producing vibrations in the external air, either directly, or indirectly through the vibrations in the sides of the pipes; and that, for the maintenance of the vibrations, a continued application of energy to the reed or the mouth-piece is necessary.

84. *Apparatus for experiments on musical strings; experiments with the monochord.*

For experiments on musical strings, and for the most convenient apparatus and manipulation, we would refer to Professor Tyndall's Lectures on Sound. The apparatus which we would recommend, as sufficient for the repetition and variation of these, is very simple and inexpensive. The basis may be a stiff piece of wood 3 or 4 inches broad and 3 or 4 feet long; to one end of this is fastened a common violin string, passing over a bridge near that end, and passing over another bridge near the other end, and then passing over a pulley and sustaining a scale-pan, in which various weights may be placed. For the experiments on harmony there should be a second string mounted in a similar way parallel to the first: one of the strings should be provided with a moveable bridge, which can be planted at any arbitrary point under it. For producing sound, a common violin-bow is to be used. Confining ourselves for the present to the single string, we may point out as the experiments most worthy of attention that, by damping the motion of the string by a touch of the finger at the middle, at $\frac{1}{3}$ length, at $\frac{1}{4}$ length, &c., still exciting the movements by the bow, pure notes will be produced which the musical ear will recognize as the harmonics of the fundamental note; and that, by putting small pieces of paper on the string, when the damping is at $\frac{1}{3}$ length or $\frac{1}{4}$ length, those which are (in the former) at $\frac{2}{3}$ length, or those which are (in the latter) at $\frac{2}{4}$ length and $\frac{3}{4}$ length will remain, showing that those points of the string are quiescent, while all others are thrown off by the vibrations. Also, by weighing the string, and ascertaining the weight in the pan, the number of vibrations per second can be found (see Article 73).

85. *Number of vibrations of air in a second for a fundamental musical note.*

This may be a convenient place for alluding to the number of vibrations corresponding to a known musical note called C of the counter-tenor scale (see Article 71). The oldest accurate information that we possess is that given by Dr Smith, Master of Trinity College, in his *Harmonics.* He determined the number by the vibrations of a string to which a weight was suspended; the vibrating part of the string being measured from its suspension-point to its point of attachment to the weight. And he says (in 1755 we believe), "The Trinity organ was now depressed a tone lower, and thereby reduced to the Roman pitch, as I judge by its agreement with that of the pitch-pipes made about 1720." His experiments, duly interpreted, give for C, about the year 1720, 465·8 vibrations per second.

The history since that time is given in a Report of the Society of Arts. It contains the following abstract.

Handel's value for C, 1740	$499\frac{1}{5}$.
A pitch recommended on grounds of theoretical convenience, as admitting of continued halving	512.
The Philharmonic Society, 1812—1842...	$518\frac{2}{5}$.
A French Commission, 1859	522.
A German Congress (Stuttgard), 1834 ...	528.
The Italian Opera, 1859	546.

The Society of Arts recommended 528 for permanent adoption, and tuning-forks made under their authority are sold at a trifling price by Messrs Cramer, guaranteed to give that number of vibrations. Theoretically, the number is convenient, as it can be halved down to 33, and can also be divided by 3. (On these advantages, see a subsequent section.)

It seems nearly impossible to prevent the continued rise of pitch. Among other causes, the desire of an ambitious singer to exhibit a voice higher in tone than can be imitated by any other, leads to the occasional raising of the pitch of all instruments: then in a short time music is written for the use of ordinary singers with reference to that raised pitch, and then the rise is established for ever.

86. *Apparatus for experiments on musical pipes, and results of some experiments.*

For experiments on pipes, we would recommend the student, having furnished himself with a tuning-fork as above mentioned, to procure some pipes which admit of alteration of length, and which can also be stopped at pleasure. A light wooden pipe $1\frac{1}{2}$ inch square and 18 inches long, open at both ends; with a tin pipe 12 inches long that will just slide in it; and with plugs for the two parts of the pipe, each plug managed by a long wire-stalk; will be found convenient. The student will be struck with the effect of a stopped pipe 6 inches or 18 inches long, or an open pipe 12 inches or 24 inches long, in resounding to a tuning-fork whose note without the pipe could scarcely be heard at all (see Articles 79 and 80). Various openings in the side of the pipe will suggest themselves. This combination also facilitates the observation of the different intensities of sound produced by the tuning-fork as it is turned into different positions, and of the vanishing of the sound in some positions (Article 49).

Among the special experiments on the vibrations of air in musical pipes, we know none so important as those by the late Mr Hopkins, published in the *Transactions of the Cambridge Philosophical Society*, Vol. v. The vibrations of air at the mouth of the pipe were produced by the vibrations of a plate of glass, vibrating (apparently) in a known time: it was found that a small distance of the glass-plate from the pipe made the phænomena the same as if the pipe had been quite open to the external air (as the experimenter who uses a tuning-fork will also find), but with particular cautions in some instances the phænomena were made identical with those which belong to a closed pipe. The examination, however, in which Mr Hopkins was most successful was the determination of nodal points; which was effected by gradually lowering into the tube a stretched membrane carrying a very small quantity of sand, and noting its place when the sand was not shaken by the air. It was thus found that the node next to the open mouth of the pipe was somewhat less distant from it than that given by theory (Article 79); or, which amounts to the same thing, that the place where the air has always the same density as the external air is not exactly at the pipe's mouth but somewhat exterior to it (as is suggested in Article 79). The experiments, generally, were experiments on resonance; and in one of these, Mr Hopkins appears to have fallen precisely on the case described in Article 81, where a very loud sound is produced. The whole of this paper, theory and experiment, is well worthy of the reader's attention.

Professor Tyndall has succeeded in exhibiting (though not with the same accuracy) the tremulous variations of density within a pipe, by placing thin membranes upon holes in the side of the pipe, and using these membranes as the bases of small gas-chambers for gas-lights; at those places where the variations of density are considerable, the agitation of the membranes extinguishes the gas-lights.

SECTION VIII.

ON THE ELEMENTS OF MUSICAL HARMONY AND MELODY, AND OF SIMPLE MUSICAL COMPOSITION.

87. *Fundamental Experiments on the Concord of the Octave, and on other principal Concords of notes sounded in succession.*

There are very few persons, possessing musical perception of the most moderate delicacy, to whom the interval of tones called the Octave (we shall hereafter give the derivation of this name) is not practically known. If a person, hearing or uttering a musical sound, endeavours to produce a sound either higher or lower, clearly and harmoniously related to the original sound, he almost infallibly utters a note whose interval from the first is an Octave. If an uneducated musician, singing with others, finds the pitch too high, he drops his voice an exact Octave. The female voice is habitually higher than the male, in general, by more than an Octave; but if a woman and a man sing the same tune together, every note of the woman's voice is exactly an Octave above the corresponding note of the man's voice.

The discovery of the mathematical connection of this musical relation with mechanical causes is ascribed to Pythagoras. The tradition is confused, but seems to leave little doubt on the general fact. Pythagoras found that, when the tone given by a vibrating string under a certain tension was noted, if the string under the same tension was stopped at half its length, the tone produced by its vibration was an Octave higher than before.

To verify this, the apparatus described in the beginning of Article 84, with two strings, should be employed. Load the scale-pans of both strings, till they give exactly the same note; then place the moveable bridge under the middle of one string, and it will give the note which the ear recognizes as an Octave higher than the other. Now in the formula of Article 73, the time of vibration is $\frac{l}{\sqrt{(gL)}}$; but the tension L is not altered by the insertion of the bridge; l however is only half what it was before. Hence, with the rise of an Octave, the time occupied by a vibration is only half what it was before, or the frequency of vibrations is double what it was before. The same effect may be produced in experiment, by leaving l unaltered but increasing L fourfold; that is (since L in Article 73 expresses the tension of the wire), by increasing the stretching-weight fourfold.

Of the physiological origin of our strong perception of this relation of sounds, we can give no account. But the perception is very definite. We have been assured by an accomplished musician that, while in musical composition some of the other concords to be shortly mentioned admit of being a little strained, the

Octaves inexorably demand the most perfect adjustment.

On sounding in succession the two notes which differ by an Octave, with whichever we begin, the effect is pleasurable. If we rise from the lower to the upper, it is animating; if we drop from the upper to the lower, it is soothing. On sounding the two notes simultaneously, a very rich effect is produced.

Remarking then the strong sense of concord conveyed to the ear by the combination of notes whose frequencies of vibrations are in the simple proportion of 1 : 2, it readily occurs to us to examine the effects of other simple proportions of frequency. The proportion 1 : 3 gives an interval (which we shall hereafter call the Twelfth) that separates the notes rather too far for accurate judgment by the ear: and the same remark applies to 1 : 4 (the Double-Octave). But the proportion 2 : 3 (for which, the length of one string must be two-thirds of the other) gives the Fifth, an interval whose concord is inferior only to that of the Octave. The proportion 2 : 4 is only a repetition of the Octave. The proportion 3 : 4 (for which, one string must have three-fourths the length of the other) gives the Fourth; and 3 : 5 (one string having three-fifths the length of the other) gives the Major Sixth; both good concords. The proportion 4 : 5 gives the Major Third, a pleasing concord. The proportion 5 : 6 gives the Minor Third, and 5 : 8 gives the Minor Sixth: both recognized as satisfactory, but only perhaps by the ear of a practised musician. Proportions with larger numbers than these are not accepted as producing useful musical results. The pleasing effect of all the relations which we have mentioned can be verified with ease on the apparatus of Article 84. The same effects may be produced, in exhibiting the several concords just mentioned, by leaving the length of the string unaltered, and by altering its stretching-weight in the proportions 4 to 9, 9 to 16, 9 to 25, 16 to 25, 25 to 36, 25 to 64, respectively; since these produce the same changes, in the value of the formula $\frac{l}{\sqrt{(gL)}}$, as are produced by diminishing l in the proportions 3 to 2, 4 to 3, &c.

88. *Experiments on the gradual formation of the Concord of notes sounded simultaneously.*

If, instead of sounding two related notes in succession, we sound them simultaneously, one person maintaining the action of the violin-bow on both strings while another adjusts the moveable bridge, whether for unison (the exact agreement of two notes) or for any of the relations above mentioned (Octave, Fifth, &c.), the phænomena observed are these. While the bridge is far from its just position, the two notes are heard separately, in rather unpleasant discord. As the bridge approaches to its just position, a rapid rattling beat is perceived, which changes to a slower softer throb, which again changes to a very slow swell and fall, (more conspicuous in approaching to unison than in approaching to the other concords), which finally disappears, leaving a most agreeable and animating concord. The sound which then reaches the ear is not like a simple note, but it gives no idea of two sounds; although when compared with either of the original notes, as sounded on a third string, it seems to be related to them.

89. *Mechanical explanation of Concords of Harmony.*

With our knowledge that every musical note implies a series of vibrations of air, following with similarity of character and at equal intervals of time (Article 71), the explanation of all these observed facts is simple. For instance, to explain the harmony in the coexistence of two sounds separated by an octave. Geometrically, we may represent the disturbance of the air produced by the lower sound as a series of waves of a certain length travelling with sound-velocity; and the disturbance produced by the upper sound as a series of waves of half the length travelling with the same sound-velocity, and therefore always holding the same relation to the series of longer waves. The union of these produces a wave more complex than either separately. If the long wave be much the larger (in the amplitude of vibrations of its particles), the result will be a modified long wave, or (musically) a modified low note; if the short wave be much the larger, the result will be a modified short wave or modified high note; in other cases it will be different from either. Algebraically, we combine $A.\sin(Bt+C)$ with $D.\sin(2Bt+E)$, and the steps are the same.

Again, to explain the harmony in the coexistence of two sounds whose interval is a Fourth (see above). Geometrically, for every three waves producing the lower note, there are four waves producing the higher note, and these produce a complicated wave, recurring with exactly the same character after every third vibration of the lower note. The 'frequency' of that recurrence is probably too slow to catch the ear; but the continued recurrence of waves, which though complicated, are precisely similar, does produce an agreeable effect. Algebraically, as the frequency is proportional to the factor of t in the expression for the disturbance, we have to combine such a quantity as $A.\sin(3Bt+C)$ with such as $D.\sin(4Bt+E)$; if D is equal to A, the sum is

$$2A.\cos\left(\frac{Bt-C+E}{2}\right).\sin\left(\frac{7Bt+C+E}{2}\right),$$

shewing that there is a note with the rapid vibration $\frac{7Bt}{2}$, whose coefficient varies according to the slower period $\frac{Bt}{2}$; with different values of A and D, the effect is more complicated.

There is one remarkable instance in which the 'frequency' of the recurrence of the complicated waves not uncommonly shows itself. Suppose that, on a piano which is accurately tuned, without striking a note which we regard as fundamental note, we strike simultaneously the Octave above that fundamental, the Twelfth above the fundamental, and the Double Octave above the fundamental. Their vibrations are twice, three times, and four times, respectively, as rapid as those of the fundamental note. The complicated wave which they produce recurs therefore in the same form, after a period exactly the same as the simple wave which would produce the fundamental note. As that note is not very far removed in the scale, we may expect to be able to perceive, amid the complication, the fundamental note; and we frequently can perceive it. This is called the Grave Harmonic.

An able musician (unacquainted with the mathematical theory) has remarked to us that, when the three notes above mentioned are sounded simultaneously, the ear always craves something more; but, if the fundamental note is sounded with them, a magnificent concord is produced, and the ear is perfectly satisfied.

90. *Geometrical representation of Concords.*

In attempting to exhibit to the eye of the student a geometrical representation of interfering waves of air, we shall indicate one difficulty for which he must specially prepare himself. If the subject had been interfering waves of water, or interfering waves of luminiferous ether, we could have exhibited them without difficulty, because the displacement of particles which produces the wave is, in great measure in the first instance, and entirely in the second instance, at right angles to the direction of the wave's motion, and is therefore at right angles to a line of abscissæ which represents either space in the wave's course (for shewing the simultaneous state of numerous particles), or time (for shewing the successive states of the same particle). But when we treat of interfering waves of air, we must treat of motions of particles which are strictly in the direction of the wave's motion. It would not be easy to exhibit these geometrically in an intelligible form. We shall therefore represent the displacements of the particles as if they were normal to the line of the wave's course, and we must beg the student always to remember that these are merely symbolical representations of displacements which are really parallel to the wave's course.

First, to represent the concord of the Octave. We have here to combine the displacements represented by the two formulæ $\sin Bt$ and $\sin (2Bt+E)$, Article 89, A and D being made $=1$, and $C=0$. It is evident that we obtain different formulæ according to the value which we assign to E. If $E=0$, we have the curve represented in Figure 14; and this law of complex undulation recurs in the same form, as long as the concord is kept up, and thus produces the pleasing effect of continual repetition of the same undulation. If $E=90°$, we have the curve of Figure 15; this exhibits a different law of complex undulation, continually repeating itself, and also pleasing. These are the extreme cases: for, if we make $E = 180°$, we fall back on the first case. Now when we strike the notes for concord, we cannot tell whether we produce the first of these, or the second, or something intermediate: but, whichever it may be, its continued repetition is very pleasing. Nevertheless, we shall shortly treat of instances in which the difference between these two complex undulations does strike the ear as very offensive.

Second, to represent the concord of the Fourth (those of the Fifth, Third, Sixth, have characters nearly similar to this). We have to combine the displacements represented by $\sin 4Bt$ and $\sin (3Bt + E)$. If $E = 0$, we have the complex curve of Figure 16; if $E = 22°.30'$ we have the complex curve of Figure 17. These are the extreme cases: if we make $E = 45°$, we again obtain Figure 16. The continued repetition of either of these produces a sound very agreeable to the ear: so also does the continued repetition of any of the intermediate combinations; and we know not, in any particular instance of striking the two notes, which of the complex undulations we have obtained. Yet, as in the instance of the Octave, we shall find cases in which the difference between these complex undulations is very disagreeable.

91. *Suggested explanation of Concords of Melody.*

On the subject of Harmony, or the agreeable consonance of simultaneous notes, we have probably said enough. But this does not strictly apply to Melody, or the agreeable relation of successive notes. This must depend on some peculiar properties of our nervous physiology. It would almost seem that there is something in our Sensorium which is put into vibration by vibrations of air, and that these vibrations subsist after the cessation of the atmospheric cause, through a time sufficiently long to be mingled with the vibrations produced by the next atmospheric disturbance, and thus to produce the effects of genuine Harmony. The expression of "sound continuing to ring in our ears" may not be so purely poetical as is usually thought.

92. *On Beats.*

We have alluded, in describing the experimental observations of Concords, to Beats. In the case of beats observed during the operation of bringing one string into unison with another, the explanation is simple. Suppose that while one string makes 100 vibrations the other makes 101 vibrations; or, which is the same thing, suppose that while there are 100 waves from one source there are superposed upon them 101 equal waves from another source. When the two waves are exactly in the same phase, one wave increases the other, and the amount of agitation produced is very great. But after 50 waves from the first source have passed, there have passed $50\frac{1}{2}$ waves from the second source: the two waves are now in opposite phases: an advance of particles of air produced by one wave is neutralized by a retreat of the same particles produced by the wave from the other source, and the particles are left absolutely at rest. And this rest continues sensibly through several waves. After this, the relative position of the two interfering waves changes, they begin to produce a real result, and after 100 waves from the first source the two waves are again united in the greatest force. The same change goes on in every successive 100 waves. Suppose that the first string under consideration produces 400 vibrations in a second of time. Then if the second string gives 101 waves for 100 of the first, their combination will produce a strong sound and a weak sound four times every second: if their tones are brought nearer, so that the second string gives 201 waves for 200 of the first, there will be a strong sound and a weak sound twice every second: if they are adjusted still more nearly so that 801 waves of the second string correspond to 800 of the first, there will be a strong and a weak sound every two seconds of time. These are the beats of notes nearly in unison.

Algebraically, we have merely to add

$$A \,.\, \sin (Bt + 2C) \text{ to } A \,.\, \sin \{(B + 2b) \,.\, t + 2D\},$$

where b is small. The sum is

$$2A \,.\, \cos (bt + D - C) \,.\, \sin (Bt + bt + D + C);$$

which exhibits a rapid vibration, that depends on $(B+b)\,t$ differing very little from Bt, and where the coefficient of vibration has the slowly varying factor $\cos(bt+D-C)$, which slowly diminishes, vanishes, rises with opposite sign, increases, diminishes, and vanishes, &c.

In Figure 18, we have represented the Beats of Imperfect Unison as produced by two waves nearly in unison; but instead of supposing the times of vibration as 100 to 101, &c., we have (for convenience) supposed them as 10 to 11.

The treatment of two sounds nearly in concord (not in unison) is somewhat different. In the union of two waves whose interval of tone is nearly an Octave, we may suppose that we combine $\sin Bt$ with $\sin(2Bt+bt+C)$, where b is very small, and where consequently $bt+C$ changes its value very slowly, going through its phases perhaps once in half a second of time, or once in a second. Now $bt+C$ occupies the place of E in the preceding investigations. Hence, when $bt+C=0°$, or $180°$, or $360°$, &c., we have for a time the complex undulation of Figure 14; when $bt+C=90°$, or $270°$, &c. we have that of Figure 15: the complex undulation is constantly shifting from one of these forms to the other, and the effect is painful to the ear.

In like manner, when the interval of the combined sounds is nearly a Fourth, and we combine $\sin 4Bt$ with $\sin(3Bt+bt+C)$, $bt+C$ takes the place of E; and whenever $bt+C$ amounts to $0°$, $45°$, $90°$, &c., the conbined undulation is represented by Figure 16; and when it amounts to 22°. 30′, 67°. 30′, &c. the combined undulation is represented by Figure 17; and the alternate change from the state of Fig. 16 to that of Fig. 17, and from the state of Fig. 17 to that of Fig. 16, and back again, &c., is very disagreeable. We have seen it described as "an angry waspish fluttering."

93. *On the doctrine of proportions of the number of vibrations in a second as representing musical intervals.*

Before entering into the subject of Musical Scale, we must expressly remind the reader, that the relation between two musical notes expressed by the terms, a Third, a Fourth, &c., does not depend on the number of vibrations per second of either note, or the numerical difference of the numbers of vibrations per second for the two notes, but on the *proportion* of the number of vibrations for one note to the number of vibrations of the other note. Thus if a note No. 1 corresponds to 480 vibrations per second, and a note No. 2 to 720 vibrations per second, and a note No. 3 to 960 vibrations per second; then, referring to the proportions and names of intervals in Article 87, it will be seen that the note No. 3 is higher than the note No. 1 by an Octave, the note No. 2 is higher than the note No. 1 by a Fifth, and the note No. 3 is higher than the note No. 2 by a Fourth. (So that the successive intervals of a Fifth and a Fourth make up an Octave.) And if we had a note No. 4 corresponding to 640 vibrations per second; then the note No. 4 is higher than No. 1 by a Fourth, and No. 3 is higher than No. 4 by a Fifth. (So that the successive intervals of a Fourth and a Fifth make up an Octave.) But the same remarks would have applied if the four numbers had been half the preceding (240, 360, 480, 320) or any multiple or submultiple of them. In fact, every expression of *interval* is an expression of *proportion* only, and is sometimes conveniently given by logarithms (see Article 93**).

As we say that the note with 640 vibrations is a Fourth above the note with 480 vibrations; so we also say that the note with 480 vibrations is a Fourth below the note with 640 vibrations; and so for all others.

93*. *On the simple Scale of Music.*

The notes in the greater part of the tunes, songs, hymns, &c., sung by persons not acquainted with artificial music, are included within the compass of an Octave. And, whatever notes are adopted within any Octave, if we adopt similar notes in the Octaves above and below it (meaning, by 'similar notes,' notes whose numbers of vibrations bear to the number of vibrations in the first note of the series the same proportion in one Octave as in the other), we infallibly secure a series of strong concords, and also give great facilities for notation. These appear to be the principal reasons which have induced mankind to use, in keyed instruments (as the Pianoforte and Organ) or in stringed instruments where no alteration is made, during musical performance, in the length of strings (as the Harp), a series of notes defined in one Octave by the concords given in Article 87, with some additions; and to repeat them in other Octaves above and below; and even to mark them with the same letters.

The letters are A, B, C, &c. to G. Apparently at some time in the history of Music, A was considered the fundamental note. But in modern Music, C is always considered the fundamental, in the same sense in which it is taken in Article 87. Using then the proportions in that Article, we have for the notes (omitting those called Minor) the following proportionate number of vibrations:—

Fundamental.	Major Third.	Fourth.	Fifth.	Major Sixth.	Octave.
1	$\frac{5}{4}$	$\frac{4}{3}$	$\frac{3}{2}$	$\frac{5}{3}$	2
C	E	F	G	A	C

These notes are sufficient for common music. (The relations of adjacent notes are not very harmonious; for instance, the proportion of E : F is 15 : 16; that of F : G is 8 : 9; but all are closely related to C, which has acquired the name of "key-note.") But, with a very small extension of musical desires, we find that other notes are required. Having sounded any note, perhaps we desire to associate with it the Third above it. We must multiply the fraction for the Third, or $\frac{5}{4}$, by the fraction for the note. This, applied to the Third and Sixth, gives $\frac{5}{4} \times \frac{5}{4}$ or $\frac{25}{16}$, and $\frac{5}{4} \times \frac{5}{3}$ or $\frac{25}{12}$, in which the numbers are too large (Art. 87); applied to the Fourth, or $\frac{4}{3}$ it produces $\frac{5}{4} \times \frac{4}{3}$, or $\frac{5}{3}$, or the Sixth; applied to the Fifth, it produces $\frac{5}{4} \times \frac{3}{2}$, or $\frac{15}{8}$, in which the numbers are not excessively large, and which falls well between the Sixth and the Octave. This proportion $\frac{15}{8}$ is therefore adopted as Seventh, with the letter b. If we desire to associate with any note its Fourth, whose factor is $\frac{4}{3}$; applied to the Fourth or $\frac{4}{3}$ it gives $\frac{16}{9}$ (which we may consider as a "flat Seventh"); applied to the Fifth or $\frac{3}{2}$ it produces $\frac{4}{3} \times \frac{3}{2}$, or $\frac{2}{1}$, or the Octave; applied to the Sixth, it produces $\frac{20}{9}$, or (as referred to the Octave) $\frac{10}{9}$ (which when we have found a note preferable for adoption as Second we may consider as a "flat Second"). If we desire to associate with any note its Fifth; the application to the Fifth gives, $\frac{9}{4}$ or (as referred to the Octave) $\frac{9}{8}$. This falls well between c and e; and the note which is an octave below it is adopted in the same place in the first Octave as Second, with the letter D. The Fifth applied to the Sixth gives $\frac{5}{2}$, or (referred to the Octave) $\frac{5}{4}$ which is the Third or e. Thus we find that only two new notes, namely $\frac{9}{8}$ and $\frac{15}{8}$, are to be inserted in our series; and it now stands thus:

Fundamental.	Second.	Major Third.	Fourth.	Fifth.	Major Sixth.	Seventh.	Octave.
1	$\frac{9}{8}$	$\frac{5}{4}$	$\frac{4}{3}$	$\frac{3}{2}$	$\frac{5}{3}$	$\frac{15}{8}$	2
C	D	E	F	G	a	b	c

The reason for the term *Octave* is now obvious. The scale which we have thus obtained is called the "Major Scale," or sometimes the "Diatonic Scale." It is universally recognized as the foundation of Music.

Sometimes the Minor Third and Minor Sixth, $\frac{6}{5}$ and $\frac{8}{5}$, are substituted for the Major Third and Major

Sixth, producing the "Minor Scale." These two new notes, though well connected together, are not well related to the other notes; and they produce a partially discordant music, of peculiar character, usually melancholy.

93**. *Systems of application of Logarithms to the expression of musical intervals.*

Professor Pole, in an essay attached to Sir F. A. Gore Ouseley's *Treatise on Harmony*, has given the logarithms of the proportions of vibrations of different notes to those of C. Those for the simple scale above are

C	D	E	F	G	a	b	c
·00000	·05115	·09691	·12494	·17609	·22185	·27300	·30103

These numbers possess the convenience of being connected with the ordinary system of logarithms, but they do not offer facility for extension. We are permitted by Sir John Herschel to explain a system proposed by him which possesses that advantage. It consists in using such a modulus that the logarithm of 2 is 1000. Thus the logarithms of the proportions of the vibrations to those of c are

C	D	E	F	G	a	b	c	d	e, &c.
0	170	322	415	585	737	907	1000	1170	1322, &c.

It is seen here that, with the exception of the figure representing a multiple of 1000, the number corresponding to each nominal letter is the same in every octave; and that, in successive octaves, the numbers increase successively by 1000. This is probably the most convenient logarithmic scale (assuming the octave-interval as the fundamental interval for music) that can be devised.

94. *Remarks on the intervals between successive notes; extension of the Scale; appropriation of numbers of vibrations and of lengths of waves to the different notes.*

If we divide the number for each note by the number for the next preceding note, we find the following series of proportions:

From C to D, $\frac{9}{8} = 1 + \frac{1}{8}$;
From D to E, $\frac{10}{9} = 1 + \frac{1}{9}$;
From E to F, $\frac{16}{15} = 1 + \frac{1}{15}$;
From F to G, $\frac{9}{8} = 1 + \frac{1}{8}$;
From G to a, $\frac{10}{9} = 1 + \frac{1}{9}$;
From a to b, $\frac{9}{8} = 1 + \frac{1}{8}$;
From b to c, $\frac{16}{15} = 1 + \frac{1}{15}$.

Considering the fractions attached to 1 as measuring the intervals of the notes, it is seen that there are two small equal intervals (E to F, and b to c); and five large intervals, nearly equal. Although the large intervals are not double of the small ones, yet in common language the larger intervals are called *tones* and the smaller *semitones;* and the Octave-scale is reputed to consist of twelve semitones. It has even been proposed, for some keyed instruments, to make all the semitones equal; the logarithm of the proportion for each semitone being $\frac{\text{logarithm of } 2}{12}$; which, on Sir John Herschel's scale, would give for the successive notes of the diatonic scale

C	D	E	F	G	A	b	c	d	e, &c.
0	167	333	417	583	750	917	1000	1167	1333, &c.

We imagine that this system would fail at every critical point of harmony.

In order to remove all denominators of fractions, and to give to each of the numbers which are associated with the notes a magnitude that represents a physical truth, we shall multiply all the numbers of Article 93* by 480. The number thus produced for C is 480. The received number of vibrations in a second of time for the counter-tenor C (see Article 85) is $528 = \frac{11}{10} \times 480$. Therefore, each of the numbers which we shall now exhibit represents the number of vibrations of air made in $\frac{10}{11}$ of a second of time, corresponding to the note to which that number is attached.

We shall take this opportunity of adding another system of numerical elements corresponding to the different notes. In Article 30 we have shewn that

$$\text{Length of wave} = \frac{\text{Velocity of wave}}{\text{Frequency of wave}},$$

where Velocity and Frequency are to be referred to the same unit of time; it is indifferent what the unit may be. On examining the numbers in Article 65, it will be seen that, at the temperatures at which it is interesting to examine musical notes, the velocity scarcely differs from 1100 English feet in a second of time, or 1000 English feet in $\frac{10}{11}$ of a second of time. Consequently, the length of wave for each note, in English feet, will be found by dividing 1000 by the number of vibrations in $\frac{10}{11}$ of a second of time.

The notes distinguished by similar letters in different parts of our range of notes are understood in all cases to bear the same relation to the fundamental note (C, c, or c, &c.) of their own part of the range; so that, for instance, the proportion of the number of vibrations in $\frac{10}{11}$ of a second of time for G to that for C is the same as that for G to that for C, or as that for g to that for c. Also the interval from C to C is an Octave, and the interval from C to c is an Octave; and so on. Using then three Octaves, we may adopt the following table as giving the Names of the Notes, the Numbers of Vibrations of Air in $\frac{10}{11}$ of a second of time, and the Lengths of the Waves of Air in English Feet.

TENOR.

G	A	B	C	D	E	F	G
180	200	225	240	270	300	320	360
5·555	5·000	4·444	4·166	3·704	3·333	3·125	2·778

COUNTER-TENOR.

A	B	C	D	E	F	G
400	450	480	540	600	640	720
2·500	2·222	2·083	1·852	1·667	1·562	1·389

TREBLE.

a	b	c	d	e	f	g
800	900	960	1080	1200	1280	1440
1·250	1·111	1·042	0·926	0·833	0·781	0·694

In special investigations, where proportions only of the numbers of vibrations of air for different notes are required, we shall divide the number of vibrations in the second line of the Table by any convenient general divisors.

95. *On cadence, and on some general principles in simple musical composition: with instances of a Ring of Eight Bells, and of the Quarter-Chimes of St Mary's Church, Cambridge.*

Perhaps we may well begin by considering the effect of the ordinary ring of eight bells: the notes of which, as usually employed for the indications of joy or triumph, are a complete octave from c to C, (or notes at the same proportional intervals), rung in the descending order. And upon analysing our sensations, they will appear to be of this kind. From the repeated close, time after time, upon the lower C, our attention is strongly drawn to that note. In the ring of the successive bells, we perceive without effort that, bell after bell, every sound has a good relation to that lower C. (Possibly the less perfect relations of B and D to C really sharpen our perception of the better relations of the other notes to C.) But the relation of each note to that which follows it, although perceptible as harmonious, is not very harmonious. Hence the ear is impressed with a certain degree of present harmony, and with the expectation of a much better harmony, which will be produced when there occurs the stroke of the bell which unites itself in strong concord with every one of the notes past. And, on hearing that bell, the ear is satisfied, and sinks into a state of rest *. This is the Cadence. After this, there is a rise

* It appears to us that these phenomena are correctly described by the poet Moore (himself no mean musician), in the following lines:

"When Memory links the tone that is gone
With the blissful tone that is still in the ear,
And Hope from a heavenly note draws on
To a note more heavenly still that is near."

The Light of the Harem.

If the reader will change the last two lines to the following, he will completely reproduce the reasoning of the text:

"And Hope from a harmony sweet draws on
To a harmony still more sweet that is near."

of an entire Octave, which is always exciting; and the descent is then repeated with the same effect.

There is, however, a circumstance which we are unable to explain. It would seem possible that, if we rang the bells upwards, from the lowest to the highest, inasmuch as each of the notes has good concord with the highest, we should derive from that series a pleasurable sensation. This, however, does not take place; the effect is unpleasant; and so strongly that (within our knowledge) the ring of the bells in ascending series is used as the alarm-signal of fire or other danger*. The difference of effects appears to depend on some unknown physiological cause.

We shall now proceed with another example, more complicated than the last, but much more simple than ordinary musical tunes; the Quarter-Chimes of St Mary's Church, Cambridge. These are universally acknowledged to be pleasing; they have been repeatedly copied for other public buildings; among others, for the Clock of the Houses of Parliament. Some years since, we were favoured by J. L. Hopkins, Esq., Organist of Trinity College, with an accurate statement of the tones of the various bells and their sequences in the chime. We have (for explanation) lowered every note by one entire tone, thus preserving the relation of the notes

* "The castle-bells with backward clang,
Sent forth th' alarum peal."

The Lay of the Last Minstrel.

unaltered, and have written them so modified in the following scheme. To each note we have attached a number, formed by dividing the numbers in Article 94 by 30: and which therefore represents the number of vibrations in $\frac{1}{30} \times \frac{10}{11}$ of a second of time, or in $\frac{1}{33}$ of a second of time.

First Quarter,
E, D, C, G;
10, 9, 8, 6;

Second Quarter,
C, E, D, G; C, D, E, C;
8, 10, 9, 6; 8, 9, 10, 8;

Third Quarter,
E, C, D, G; C, D, E, C; E, D, C, G;
10, 8, 9, 6; 8, 9, 10, 8; 10, 9, 8, 6;

Fourth Quarter.
C, E, D, G; C, D, E, C; E, C, D, G; G, D, E, C;
8, 10, 9, 6; 8, 9, 10, 8; 10, 8, 9, 6; 6, 9, 10, 8:

Hour Bell ... ... C.
4.

The chime for the First Quarter presents little for remark: it is a simple descending succession, giving two imperfect harmonies (10 : 9 and 9 : 8), but ending with a Fourth (8 : 6 equal to 4 : 3), a good descending harmony: every note harmonizes well with the last. For appreciating all the quatrains of the other Quarters, it must be borne in mind that the ear becomes fatigued by continual harmonious relations: and it is found necessary, from time to time, to interpose unharmonic sequences of notes. This is done in the middle of every quatrain. Thus, in the middles of the two quatrains of Second Quarter there are 10 : 9 and 9 : 10: in the middles of the three quatrains of Third Quarter there are 8 : 9, 9 : 10, and 9 : 8; and so for the four quatrains of Fourth Quarter. Each of these strikes the ear as a little hitch or dislocation, which makes the strong concord that follows very welcome. The first two notes of a quatrain present good concords in five instances and inharmonious sequences in four instances: but in every quatrain the last two notes present strong concords; and it is at the end of the quatrain that they are most desired by the ear. Finally, the sound drops by an entire Octave upon the Hour Bell: and this last sequence is eminently satisfactory.

Perhaps there could be no better education for a young Cambridge musician, than to learn habitually to associate the tones of the several bells with the numbers that we have attached to them, and to repeat those numbers on hearing the sounds of the bells.

96. *Instances of musical melodies: "God save the Queen," and "Adestes Fideles."*

In the following exhibition of "God save the Queen," as sung by a single voice, the first line gives the words, the second line shews (by figures) the proportional time occupied by each syllable, the third gives the letter of the musical note, and the fourth contains the number of vibrations in the scale at the end of Article 95 divided by 10, representing therefore the number of vibrations in $\frac{1}{11}$ of a second of time.

God save our gra-cious Queen Long live our no-ble Queen
2 2 2 3 1 2 2 2 2 3 1 2
C C D B C D E E F E D C
48 48 54 45 48 54 60 60 64 60 54 48

God save the Queen | 4 | Send her vic-to-ri-ous Hap-py and
2 2 2 2 | 4 | 2 2 2 3 1 2 2 2 2
D C B C | | G G G G F E F F F
54 48 45 48 | | 72 72 72 72 64 60 64 64 64

Glo-ri-ous Long to reign o-ver us G-o-d save the Queen.
3 1 2 2 1 1 1 1 3 1 1 1 ½ ½ 2 2 2
F E D E F E D C E F G a G F E D C
64 60 54 60 64 60 54 48 60 64 72 80 72 64 60 54 48

The whole strain is included within the limits of an Octave; the lowest note being B and the highest a.

The conspicuous repetition of the note C at the beginning of the melody leaves no doubt on the ear that C is the key-note to which all the others are to be referred: and accordingly all the other notes are in the diatonic scale. And the final cadence, dropping upon C by a series of notes which bear to C the proportions

$\frac{5}{3}$, $\frac{3}{2}$, $\frac{4}{3}$, $\frac{5}{4}$, $\frac{9}{8}$, is very satisfactory, (the feebleness of $\frac{9}{8}$ being however perceptible to the ear), and produces an excellent musical effect.

Now if we examine the relations of successive notes all through the strain, we shall find that the only decided harmonic relations are the following. From "our" to the first syllable of "gracious", the proportion is 6 : 5; from "Queen" to "Send" the proportion is 2 : 3, and this rise of a Fifth produces an animating effect; between "reign" and the first syllable of "over" there is the proportion 4 : 5. All the other relations of adjacent notes are unharmonic. On the whole, the music is meagre, if considered only with reference to the relations of adjoining notes. But, considered with reference to its phrases; the first line of the song may be accepted as entirely in the vibrations 48, with small variations to prevent monotony; and the second entirely in the vibrations 60, with similar variations: and the relation of these or 4 : 5 is markedly harmonic. The third line is mainly in 48, the fourth in 72 or $48 \times \frac{3}{2}$, the fifth in 64 or $48 \times \frac{4}{3}$. The remainder, with the exception of the first note of "God", 80 or $48 \times \frac{5}{3}$, is less distinctly marked (we believe that the melody is not easily learnt with accuracy by uneducated persons); but it is in great measure redeemed by its final cadence. In practice it is usually accompanied with instrumental music of rich chords (to be mentioned hereafter).

Of the melodies delivered to us by the religion of former ages, perhaps that known by its commencing words, "Adeste fideles", or by its more usual name, "The Portuguese", is the most magnificent. It is very easily learnt with accuracy by uneducated persons. We will treat it in the same manner as the last.

Ad-es-te fi-de-les læ-ti tri-um-phan-tes, ve-ni-te ve-ni-te
1 2 1 1 2 2 1 1 1 1 2 1 1 2 1 1 1 1
C C G C D G E D E F E D C C B A B C
48 48 36 48 54 36 60 54 60 64 60 54 48 48 45 40 45 48

in Beth-le-hem. Na-tum vi-de-te re-gem an-ge-lo-rum, ve-ni-te
1 1 2 2 4 2 1 1 2 2 1 1 1 1 2 1 1 1 1
D E B A G G F E F E D E C D B G C C B
54 60 45 40 36 72 64 60 64 60 54 60 48 54 45 36 48 48 45

ad-o-re-mus, ve-ni-te ad-o-re-mus, ve-ni-te ad-o-re-mus
1 1 2 1 1 1 1 1 1 2 1 1 1 1 1 1 2 1 1
C D C G E E D E F E D E F E D C B C F
48 54 48 36 60 60 54 60 64 60 54 60 64 60 54 48 45 48 64

Do-mi-num.
2 2 3
E D C
60 54 48

The figures defining the time occupied by each syllable have not necessarily the same value as in the analysis of the former melody; but those which are related to the vibrations corresponding to the musical notes have the same value in both.

The lowest note is G; the highest (which occurs but once) is G; and the range of the melody is exactly an Octave.

In this melody, as in the other, the frequency of C in the beginning fixes it in the ear as the key-note. All the notes are related to it, in the diatonic scale. The final cadence upon C, though not consisting of so many steps as in the former instance, has the descent through $\frac{4}{3}$, $\frac{5}{4}$, $\frac{9}{8}$, which is sufficient and effective. There is an excellent imperfect cadence of $\frac{5}{3}$, $\frac{5}{4}$, $\frac{10}{9}$, upon the last syllable of "Bethlehem," which is a Fourth below the key-note. If we examine the relations of the successive notes of the melody, we find that there are in all 58 relations, of which 13 are strongly harmonic, being expressed by $\frac{1}{2}$, $\frac{2}{3}$, $\frac{3}{4}$, $\frac{3}{5}$, $\frac{4}{5}$, or their reciprocals. It would be wearisome to examine the fitness of each of these to its place; but there is one which deserves especial attention. From the last syllable of "Bethlehem" to the first syllable of "Natum" the interval is an entire Octave upwards. This animating rise, leading to the word which is emphatically and characteristically the subject of the poem, has a most splendid effect.

The grandeur of this piece of music, and the facility with which it is apprehended by mankind in general, appear to be fully explained by the consideration of the numerical relations of the vibrations corresponding to the notes.

97. *On Enriched Music, and Singing in Parts.*

The melodies that we have considered, if played upon an instrument which does not necessarily at every touch or blast produce a combination of several harmonic notes, would seem too simple. It is necessary to ornament them by combining, with almost every note in the melody, a series of related notes, to be struck simultaneously with it. They are usually lower notes, in harmonic relation. The combined system, of the original note and these related notes, constitutes a "chord."

As instances of chords, we will exhibit (in the same form as in the last Article) the first part of "God save the Queen," and the second part of "Adeste fideles."

God save our gracious Queen Long live our noble Queen God save the Queen

2	2	1	1	3	1	2	2	2	2	3	1	2	2	2	2	4
C	C	D	C	B	C	D	E	E	F	E	D	C	D	C	B	C
G	G	A		G	A	B	C	C	D	C	F	G	F	G	F	G
E	E			D			G	G	A	G		E		E	D	E
48	48	54	48	45	48	54	60	60	64	60	54	48	54	48	45	48
36	36	40		36	40	45	48	48	54	48	32	36	32	36	32	36
30	30			27			36	36	40	40		30		30	27	30

Natum		videte			regem		angelorum				Venite			adoremus			
2	1	1	2	2	1	1	1	1	2	1	1	1	1	1	1	2	1
G	F	E	F	E	D	E	C	D	B	G	C	C	B	C	D	C	G
C	B	C	D	C	B	C			A	G		G	G	G	G	B	
72	64	60	64	60	54	60	48	54	45	36	48	48	45	48	54	48	36
48	45	48	54	48	45	48		40	36		36	36	36	36	45		

Venite			adoremus				Venite			adoremus				Dominum.			
1	1	1	1	1	2	1	1	1	1	1	1	2	1	1	2	2	3
E	E	D	E	F	E	D	E	F	E	D	C	B	C	F	E	D	C
C	C		C	B	C	B	C	B	C	G	G	G	G	C	C	B	
60	60	54	60	64	60	54	60	64	60	54	48	45	48	64	60	54	48
48	48		48	45	48	45	48	45	48	36	36	36	36	48	48	45	

In the greater part of these chords, the proportions of the numbers of vibrations admit of being expressed by very low numbers; for instance, the proportion $\begin{matrix}48\\36\\30\end{matrix}$ is the same as $\begin{matrix}8\\6\\5\end{matrix}$; both $\begin{matrix}45\\36\\27\end{matrix}$ and $\begin{matrix}60\\48\\36\end{matrix}$ are the same as $\begin{matrix}5\\4\\3\end{matrix}$; and so for others. On the first syllable of "Natum" in the second piece of music (to which syllable we have called attention before), the chord $\begin{matrix}72\\48\end{matrix}$ is the same as $\begin{matrix}3\\2\end{matrix}$, or two notes at the interval of a Fifth, a most powerful harmony. There are several chords containing

D = 54, as $\frac{54}{40}$, $\begin{matrix}64\\54\\40\end{matrix}$, $\frac{54}{32}$, which cannot be accepted as harmonious; if, however, D be made = 56 (which, giving a relation $\frac{7}{6}$ to the key note C, appears permissible), the proportions become $\frac{7}{5}$, $\begin{matrix}8\\7\\5\end{matrix}$, $\frac{7}{4}$, and the harmony is good. This is an instance of what frequently occurs in musical composition, that the harmonious relation of two adjacent or simultaneous notes is more important than the harmonious relation of either note to the key-note. The chords $\begin{matrix}45\\32\\27\end{matrix}$ or (as corrected) $\begin{matrix}45\\32\\28\end{matrix}$, $\frac{56}{45}$, $\frac{64}{45}$, appear to be incorrigibly discordant (though $\frac{54}{45}$, the original form of $\frac{56}{45}$, is equivalent to $\frac{6}{5}$, and is harmonic).

Singing in Parts may be understood as an exact imitation by different human voices of that which is done here by different fingers upon an instrument. The highest voice usually sings the notes as given in Article 96, or as in the upper line of the harmonized music just exhibited; the other voices at the same time sing notes represented by those of the second and third lines of the harmonized music. The skill of a composer of Music in Parts is shewn by his assigning, to those other voices, notes which will produce good harmony with each note of the highest voice, and which at the same time will produce for each person, considered alone, a species of tune that can be followed correctly with ease.

98. *On the necessity for the supplementary notes called "flats" and "sharps."*

We have exhibited to the reader two pieces of music in which every sound is expressed by a note of the diatonic scale. And the reason of our having been able sufficiently to express all by that scale is, that we have taken C for the key-note of our melodies; and that, in these special instances, that pitch has appeared so well suited to the subject that there has been no temptation to adopt any other key.

But it will easily be imagined that we cannot bind ourselves to this condition. We may desire to make our tune somewhat higher or somewhat lower, but we cannot venture to raise it an entire octave so as to have c for key-note, or to depress it by an entire octave so as to have C for key-note. And we may now consider how the power of using any other note as key-note can be obtained.

Proceeding upwards from C we have in the diatonic scale, two tones, then one semitone, then three tones, then one semitone. We desire to have a similar succession of intervals for any other key. Suppose, for instance, we adopt D as key-note.

Diatonic Scale proceeding from C	C	D	E	F	G	a	b	c	d	e	f
Scale with similar intervals proceeding from D		1	2		3	4	5	6	7	1′	2′

Here, the notes 1, 2, 4, 5, 6 on our new scale may be represented by D, E, G, a, b, on the old scale. The interval from 1 to 2 will not be precisely the same as that from C to D, and so in other parts of the scale; but they will be sufficiently near to them to pass (in ordinary estimation) for similar intervals, perhaps slightly altering the character of the music. But the notes 3, 7, on our new scale, have no representative on the old scale; and, for representing them, it is necessary to introduce a note nearly midway between F and G, and a note nearly midway between c and d (and similarly a note between C and D). The note between F and G is called indifferently F sharp, F♯, or G flat, G♭; and that between C and D is called C♯ or D♭.

Now when we adopt other notes as key-notes, we find that we must interpose notes between various notes of the original diatonic scale; and ultimately we find that in every interval of a complete tone it is necessary to insert a new note. The whole scale, for one octave, then becomes this:

	C♯		D♯			F♯		G♯		a♯	
C		D		E	F		G		a		b c.
	D♭		E♭			G♭		a♭		b♭	

This is sometimes called the chromatic scale.

The new notes are those corresponding to the black keys of a pianoforte.

In some theoretical works, it has been proposed to consider D♯ and E♭, and other pairs similarly related, as slightly different notes; as will be seen below.

When B, or C♯, or F♭, is adopted as key-note, every one of the five new notes is used. With key-note C♯, very good music may be performed on the black keys only; but it will be found that it wants the Third interval (four semitones) and the Seventh interval (eleven semitones). With key-note F♯ also, the black keys are sufficient; but the Fourth interval (five semitones) and the Seventh interval (eleven semitones) are wanting; this is frequently considered as a characteristic of Scotch music, which, consequently, can be played on the black keys only, if the key-note is F♯.

If we examine theoretically the tuning which can be advantageously adopted for the flats and sharps, we are led to considerations like the following:

Proceeding from C = 240;

we may take for C♯ or D♭

$$256 = 240 \times \frac{16}{15} = 240 \times \frac{4}{3} \times \frac{4}{5} \text{ or } 240 \times \frac{4}{3} \div \frac{5}{4}$$

(a Third below the Fourth).

D = 270.

Then for D♯ or E♭ we may use either $280 = 240 \times \frac{7}{6}$ (which will associate well with F♯); or

$288 = 240 \times \frac{6}{5}$ (which will associate with $\frac{7}{5}$, $\frac{8}{5}$, and $\frac{9}{5}$).

E = 300.

F = 320.

For F♯ or G♭ we may adopt $336 = 240 \times \frac{7}{5}$. It will be seen however in Article 100 that, for reference to other keys, $\frac{675}{2}$ is preferable.

G = 360.

For G♯ or A♭ there are the competing claims of $375 = 240 \times \frac{5}{4} \times \frac{5}{4}$, a Third interval above the Third; and $384 = 240 \times \frac{8}{5}$ (associating with $\frac{6}{5}$ and $\frac{7}{5}$).

A = 400.

For A♯ or B♭, the competing values are $420 = 240 \times \frac{7}{4}$ (which associates with $\frac{7}{6}$ and $\frac{7}{5}$); and $432 = 240 \times \frac{9}{5}$ (associating with $\frac{6}{5}$, $\frac{7}{5}$, $\frac{8}{5}$).

B = 450.

C = 480.

We apprehend that, for music of the highest order, it may be necessary for D♯ or E♭, for G♯ or A♭, for A♯ or B♭, sometimes to use one of the values which we have indicated, sometimes the other; or, in fact, to consider D♯ and E♭, G♯ and A♭, A♯ and B♭, as different notes.

SECTION IX.

ON INSTRUMENTAL MUSIC, AND THE ADAPTATIONS OF MUSIC REQUIRED BY SPECIAL INSTRUMENTS.

99. *On the characteristic differences of construction, between the Organ, the Pianoforte, and the Violin.*

Among the great variety of musical instruments, we have selected three, as embodying the most important differences of construction. We shall describe them briefly, but sufficiently perhaps to convey an idea of the bearing of these differences upon their musical powers.

The organ is to be conceived as essentially an instrument of three dimensions. For perspicuity of language, we will consider the horizontal line along the front as x, the horizontal line normal to the front as y, and the vertical line as z. Then the parallelopiped may be understood as being filled with vertical pipes all parallel to z. The front row, parallel to x, for which y nearly $= 0$, are pipes with plain mouth-pieces, corresponding to the notes and interpolated notes in Article 98. The next row parallel to x, for which y has a uniform value, consists of pipes having a uniform definite relation to the front row, either in tone or in construction; for instance, the note of every pipe may be higher than that of the corresponding front pipe by a Fifth; or every pipe may have a reed mouth-piece instead of a plain mouth-piece. The third row, with another value of *y*, consists of pipes with some other variation, but still such that the note of every one of these pipes bears the same relation to the note of the corresponding pipe in the front row. And so on for many rows. And thus, if we consider any row of pipes in the direction *x*, all are of similar construction and give notes related as in the chromatic scale; but if we consider any row in the direction of *y*, these are pipes giving different harmonical notes and differently constructed, the same description applying to every row parallel to *y*, except that each is dependent on the note of its front pipe.

All the pipes in any one row in the direction of *x* can be stopped at once by the slider called a "stop." All the pipes in any one row in the direction of *y* are closed unless opened by a key (like that of a piano-forte). To open any individual pipe and allow the air from the wind-chest to act in it, the proper stop must be open, and the finger of the musician must press the key; the stop and the key being those which correspond to the *y* and *x* whose intersection is at the pipe in question.

From this it will be seen that the instrument possesses in itself the power of giving chords, and even of giving harmonic chords produced by mouth-pieces of different kinds uttering sounds of different qualities, which, however, by the mechanism of the "stops," are at the command of the musician. But he has no power of altering the pitch of the note which he plays; he must accept that as it was left when the organ was tuned.

The pianoforte is known as a stringed instrument; but few persons at first view imagine what a number of strings it contains. A good piano has about 75 keys, and each of these works a hammer that strikes three strings (tuned to the same note) simultaneously; so that there are more than 200 strings to be adjusted in pitch. The tuner begins by screwing up one triplet of strings till they give the same note as a tuning-fork which sounds the note C (528 vibrations per second), or sometimes A; from this he, by ear, adjusts the triplets for the other notes of that scale, included within the compass of an Octave; and from each of these notes he steps upwards and downwards, by ear, through entire Octave-intervals, to form all the intermediate notes of the higher and lower octave-scales. In playing on the piano, the striking of one key does not, as in the organ, produce any number of harmonic sounds (except on the principle of Article 76); to produce these, it is necessary for the musician to strike simultaneously different keys with different fingers. In the impossibility of having the pitch altered by the player during the performance, the piano resembles the organ. If there are any faults or peculiarities in the tuning, the player has no control over them.

Most of the remarks on the piano apply also to the harp.

The violin is an instrument of a totally different class. It has only four strings, which are stretched over a curved bridge; and under no circumstance can the bow touch more than two of these, and generally it touches only one; so that the instrument possesses little power of producing harmonic notes, except for the reasons given in Article 76. But the instrument is so held in the left hand that, by application of the four fingers, either of the strings can be defined in length (by pressing the string upon a part of the fixed bar which is nearly parallel to the strings), and thus the tone of the string can be varied at the moment, through a range sufficient to exhibit any note between the fundamental note of one string and the fundamental note of the next string, and with an accuracy depending on the precision of the player's perceptions and the delicacy of his finger-action. We imagine that a skilful player perceives the tone given by the string long before it is heard by the bystander, and instantly adjusts his pressure to make the intended concord perfect. With this presumed power, the violin is far more accurate for the production of exact melody than the organ or the pianoforte or harp.

The fundamental notes to which the different strings are adjusted are the following (adopting, as representing the number of vibrations in $\frac{10}{11}$ of a second of time, the numbers in Article 94):

For the four strings of the violin	E	A	D	G
	600	400	270	180
For the four strings of the viola	A	D	G	C
	400	270	180	120
For the four strings of the violoncello	A	D	G	C
	200	135	90	60

It will be seen here that two of the intervals on each instrument are Fifths $\left(\text{proportion of vibrations } \frac{2}{3}\right)$; and that the other interval is $\frac{270}{400}$, or $\frac{270}{405} \times \frac{81}{80}$, or $\frac{2}{3} \times \left(1 + \frac{1}{80}\right)$, differing from a perfect Fifth only by the interval numerically expressed by $\frac{1}{80}$. This small interval is called a "Comma"; it is the smallest which is recognized in ordinary music; it is held to be perceptible in all cases to the ear, and very offensive when it deranges the numerical accuracy of a chord. In practice, the intervals of the violin strings are in the first instance made perfect Fifths (A being taken from the tuning-fork or piano); and the notes of the other strings will therefore depart a little from those of the diatonic scale. This will lead us to the considerations of the next Article.

100. *On the necessity for Temperament, in reference to chords and harmonies within the compass of an Octave.*

Of all parts of theoretical music, the theory of Temperament is the most troublesome. We cannot pretend here to go into it to the extent which would be useful to the professional musician. All that we can do is, so to explain the state of the matter that the student may be able to understand the nature of the evils of which musicians complain; and that, if he should desire to examine further the plans suggested by professional musicians, he may find himself prepared with the numerical bases on which all must depend.

The difficulties in question attach only to keyed or other instruments whose tones cannot be altered by the musician in the act of playing. They apply therefore to the two most important of all, the organ and the pianoforte; and they are perhaps more important for the organ; because, as the key is there held down so as to give an enduring sound, any fault in the harmonies is more obvious to the ear than with the pianoforte, where the sound is produced by an impact on the strings, and is damped as soon as possible.

We shall examine, first, the evils produced in the chords, by the use of different notes as key-notes; which evils we shall consider to be represented by the departure of the proportions of vibrations in the 1st, 2nd, &c. notes on that scale from the proportions of vibrations in the 1st, 2nd, &c. notes in the accurate diatonic scale; secondly, the difficulties in the prolongation of long series of harmonies, and in the combination of instruments like the piano with instruments like the violin.

For understanding more easily the following investigation, we would recommend the student to provide himself with a scale on paper including two octaves of the keys of a piano as shewn in Article 98; thus

A♯ C♯ D♯ F♯ G♯ A♯ C♯ D♯ F♯ G♯

A B C D E F G A B C D E F G a

B♭ D♭ E♭ G♭ A♭ B♭ D♭ E♭ G♭ a♭

and also a corresponding moveable scale of card-board with notes 1, 2, &c. to 8, at intervals representing tones and semitones in the diatonic scale beginning with C; thus

1 2 3 4 5 6 7 8

then, for ascertaining the notes which will be used in conjunction with any assigned key-note, it is only necessary to apply the card-board scale to the paper scale so that figure 1 is opposite to the key-note, and the other notes to be used will then be those opposite to the figures 2, 3, 4, &c.

Upon examining a great quantity of piano-music, we find that the note E♭ is used for key-note twice as often as any other; and that those next to it in general favour are F, A♭, C, G, D, D♭. We will examine successively the scales of notes for these seven key-notes. As the scale for E♭ will require three flats, we shall select from the values suggested in Article 98 the three values which evidently give the best harmonies in this scale, and shall retain those values in the succeeding scales; and, as need may arise for values for new flats and sharps in the succeeding scales, we shall adopt and retain them in the same manner. Below each note, we shall place its number of vibrations in $\frac{1}{11}$ of a second of time, and the proportion which it bears to the key-note of the scale, and (where necessary) the factor connecting this proportion with the proportion which ought to exist on the diatonic scale.

1	2	3	4	5	6	7	8

Key-note E♭. We have

E♭	F	G	A♭	B♭	C	D	E♭
	320	360			480	540	

and we have to assign vibrations to E♭, A♭, B♭, from the table in Article 98. It will be easily seen that 288 for E♭ will harmonize perfectly with G, C, D, and very well with F. Adopting it, it will then be seen that 384 for A♭ and 432 for B♭ harmonize perfectly with it. Thus we have

E♭	F	G	A♭	B♭	C	D	E♭
288	320	360	384	432	480	540	576
1	$\frac{10}{9}$	$\frac{5}{4}$	$\frac{4}{3}$	$\frac{3}{2}$	$\frac{5}{3}$	$\frac{15}{8}$	2
	$\left(=\frac{9}{8}\times\frac{80}{81}\right)$						

Key-note F. We have

F	G	A	B♭	C	D	E	F
320	360	400	432	480	540	600	640

We have no need to supply any new number. The proportions are

1	$\frac{9}{8}$	$\frac{5}{4}$	$\frac{27}{20}$	$\frac{3}{2}$	$\frac{27}{16}$	$\frac{15}{8}$	2
			$\left(=\frac{4}{3}\times\frac{81}{80}\right)$		$\left(=\frac{5}{3}\times\frac{81}{80}\right)$		

Proceeding in the same manner, we find that we have to supply a value of D♭ for key-note A♭, a value of F♯ for key-note G, a value of C♯ or D♭ for key-note D, and a value of G♭ or F♯ for key-note D♭. Thus we obtain

Key-note A♭.

A♭	B♭	C	D♭	E♭	F	G	a♭
384	432	480	512	576	640	720	768
	$\frac{9}{8}$	$\frac{5}{4}$	$\frac{4}{3}$	$\frac{3}{2}$	$\frac{5}{3}$	$\frac{15}{8}$	2

Key-note C.

It is unnecessary to examine this scale, as it is the diatonic scale with which the others are compared.

Key-note G.

G	A	B	C	D	E	F♯	G
360	400	450	480	540	600	675	720
	$\frac{10}{9}$	$\frac{5}{4}$	$\frac{4}{3}$	$\frac{3}{2}$	$\frac{5}{3}$	$\frac{15}{8}$	2
	$\left(=\frac{9}{8}\times\frac{80}{81}\right)$						

Key-note D.

D	E	F♯	G	A	B	C♯	D
270	300	$337\frac{1}{2}$	360	400	450	512	540
	$\frac{10}{9}$	$\frac{5}{4}$	$\frac{4}{3}$	$\frac{40}{27}$	$\frac{5}{3}$	$\frac{256}{135}$	2
	$\left(=\frac{9}{8}\times\frac{80}{81}\right)$			$\left(=\frac{3}{2}\times\frac{80}{81}\right)$		$\left(=\frac{15}{8}\times\frac{2048}{2025}\right)$	
						$\left(=\frac{15}{8}\times\frac{89}{88}\text{ nearly}\right)$	

Key-note D♭.

D♭	E♭	F	G♭	A♭	B♭	C	D♭
256	288	320	336	384	432	480	512
	$\frac{9}{8}$	$\frac{5}{4}$	$\frac{21}{16}$	$\frac{3}{2}$	$\frac{27}{16}$	$\frac{15}{8}$	2
			$\left(=\frac{4}{3}\times\frac{63}{64}\right)$		$\left(=\frac{5}{3}\times\frac{81}{80}\right)$		

In two of these scales, the diatonic character is perfect; and in two, a single note only is in error.

The principal difficulty is with the Fifths. Proceeding upwards from the key-note to the note 5, the Fifths require correction in only one scale (key-note D); but proceeding upwards from note 5 to note 9, whose vibrations will be double those of note 2, we find that with key-notes E♭ and G the proportion is too small, and with key-note D it has the just magnitude, $\left(\frac{3}{2}\right)$, only because the first proportion from D to A $\left(\frac{3}{2}\times\frac{80}{81}\right)$ is too small. With key-note F, the proportion of note 4 to 1 is too large, but the proportion of note 6 to 1 is equally too large, so that the proportion of note 6 to 4, $\left(\frac{5}{4}\right)$, is just, and their harmony perfect. Note 7 fails in one instance, and cannot be reconciled with either of those which ought to be its bases, notes 3 and 5.

Innumerable attempts have been made to alter some of these notes, so as to produce the most perfect harmony where it is most wanted; and this alteration is called *Temperament.* But it is easily understood that, where the errors for different scales, as affected by individual notes, are taken into one view, the Temperament which improves one scale will injure others. Thus if F be altered to correct the scale with key-note E♭, it will ruin that with key-note F, and will injure others. An attempt is usually made to throw as much as possible of the discord on one particular scale; and some assemblage of notes in that scale which ought to form a chord (as in Article 97) and which really produces a discordant sound, very unpleasant to the ear, is called the *wolf.*

It appears to us that a scale such as we have proposed will answer well for the most valuable keys; especially if the compositor, in writing the music, would sometimes avoid those discordances of successive notes in the music, which would (though rarely) be produced by incautious use of the notes. Probably a similar principle of selection might be adopted for organ-music;

taking first that key in which the greatest quantity of music or the most important music is written, and supplying the best values of flats and sharps for it, as we have done in piano-music for E♭; and then proceeding with the key of next importance, &c.

101. *On the necessity for Temperament, in reference to long series of harmonics, and to the combination of different instruments.*

The concord of notes which is most powerfully felt by the ear, as we have stated in Article 87, is that of two notes at the interval of an Octave. But this interval is so large that it is rarely repeated in musical compositions. The next degree of concord is that of two notes at the interval of a Fifth (proportion 2 : 3), and their interval is so much smaller that several repetitions (one such interval rising above another, or one falling below another) may be occasionally introduced.

Now if we begin (as the most advantageous place) from one of the notes marked F, as **F** = 160, we have the following succession of perfect Fifths (the proportion of vibrations being always 2 : 3):

From 160 to 240, **F** to C;
From 240 to 360, C to G;
From 360 to 540, G to D;
From 540 to 810, D to a note beyond a,

$$\left(\text{vibrations} = \frac{81}{80} \text{ of those of a}\right).$$

Thus we find that, even beginning at the most advantageous place, we cannot have more than three consecutive perfect Fifths; and, generally speaking, we cannot have so many. If we could put up with an error represented by the factor $\frac{81}{80}$ (either in the last interval, or distributed through the three intervals) we might proceed further. It will be seen that we must either diminish the Fifth-interval a little below $\frac{3}{2}$ ("flatten the Fifths") or increase the Octave-interval above 2 ("sharpen the Octave"); no compositor for the piano dares to recommend the latter course. We are inclined to think that it is best also to leave the Fifths unaltered.

In the use of the violin, however, the temptation to retain the perfect Fifths is very strong. In the process of tuning, every interval of the fundamental notes of the strings is made a perfect Fifth; and, if the violin only were used, it can scarcely be doubted that they would be retained in that state. But the number of pianos in society is so great that it is necessary to make the violins yield to them; and the violins accordingly are subject to the Temperament of having their Fifths a little flattened. Their different strings can then be made to harmonize pretty well with the notes in different Octaves of the piano.

Section X.

ON THE HUMAN ORGANS OF SPEECH AND HEARING.

102. *On the human organs for producing musical notes.*

A tolerably clear idea was formed by anatomists, many years ago, of the nature and action of the organization in the human throat by which musical notes are produced. Several points of explanation, however, were wanting; these were supplied, perhaps finally, by Professor Willis, in a paper "On the Mechanism of the Larynx," published in the fourth volume of the *Transactions of the Cambridge Philosophical Society*, which may be regarded as a model of scientific and anatomical inquiry.

The top of the windpipe is closed by an apparatus which leaves for the passage of air only a long narrow slit in the back-and-front direction, called the "glottis." The sides of this slit are not solid masses of animal matter, but elastic bands or ligaments, which, though not very deep in the vertical direction, can vibrate in the right-and-left direction. Their extremities are governed in position and tension by various muscles, which have been most accurately described. The state of these elastic bands, frequently called the "vocal ligaments," under various circumstances, is as follows:—

In the ordinary state of ease, the vocal ligaments are not stretched longitudinally with any special force, and the ends of the right ligament and those of the left ligament are not pressed together. There are, in fact, special muscles for separating them, which in the state of personal ease appear to be in action, effecting that separation. The opening is then sufficiently wide to allow the breath to pass very freely; and the ligaments, in their unstretched state, will not vibrate.

For producing sound, other muscles are brought into play, namely, 1st, muscles which press together (but probably not close together) the ends of the two ligaments: 2nd, muscles which extend each ligament to any arbitrary degree of tension.

In this state, when air is forced from the lungs through the glottis, necessarily passing with great rapidity (as the chink is now very narrow), it puts the ligaments into vibration, sufficiently rapid to produce a musical note. The pitch of the precise note produced will depend on the tension given to the ligaments. So that, for utterance of a musical sound, two systems of muscular action are required. One, consequent on the volition "to utter a musical sound," is that of drawing close together the right ligament and the left ligament; the other, consequent on the volition "to utter a musical sound of a particular pitch," is that of stretching the ligaments to a definite tension.

This action may be imitated experimentally in various ways, of which we quote, as probably the easiest and best, that given in Professor Tyndall's excellent book on "*Sound, a Course of Eight Lectures, &c.*" "Roll round the end of a glass tube a strip of thin india-rubber, leaving about an inch of the substance projecting beyond the end of the tube. Taking two opposite portions of the projecting india-rubber in the fingers, and stretching it, a slit is formed; the blowing through which [by means of blowing through the tube] produces a musical sound, which varies in pitch, as the sides of the slit vary in tension."

The vibrations of the vocal ligaments produce vibrations of the air in the resonant cavity of the mouth, in the same manner in which the vibrations of a reed produce vibrations in a resonant organ-pipe (Article 81).

The compass of a singer's voice, or the interval (upon the musical scales) between the lowest note and the highest note which a singer can distinctly sing, in general does not much exceed two octaves.

The phænomena of *whistling*, we believe, have not been carefully examined. They appear, however, to correspond nearly to those of the production of musical sounds in the throat. The lips are brought, by their appropriate muscles, into a state of tension analogous to that of the vocal ligaments; and the passage of a current of air between them compels their surfaces to vibrate and to produce vibrations in the air. The resonant cavity is the hollow of the mouth; and the mechanical conditions for whistling differ from those for speaking, in this respect, that the current of air for whistling passes *from* the resonant cavity *to* the vibrating lips. Whistlers remark that, for producing their lowest notes, the tongue is very far retracted; the lower lip also appears to be somewhat retracted. The range of notes in whistling is greater than that in singing; it extends nearly to three octaves.

103. *Experiments and theory on the production and maintenance of vibrations similar to those of the human organs of sound.*

Professor Willis has pointed out very clearly the mechanical reasons which shew that there must be, in the stretched ligaments of the glottis, a tendency to produce the requisite vibrations. Thus, if we have two small frames on which laminæ of sheet india-rubber are stretched, leaving their upper edges free, as in Figure 19, and if we place these frames nearly vertical, upon a pedestal through which air can be blown upwards from below, passing between the two laminæ; then if, as in Figure 20, the upper edges are brought near together, the blast of air will force them outwards; if, as in Figure 21, they are opened wider, the blast of air will suck them inwards; if they are opened to the state in which neither of these tendencies is discoverable, then with every displacement outwards there will be a proportionate force tending to bring each towards the middle of the chink, and with every displacement inwards there will be a proportionate force tending to push each from the middle of the chink. So that each lamina, considered alone, is, upon receiving any displacement from a certain position of rest, subjected to a force tending to bring it to that position of rest. This is the law of force proper for maintaining vibration (it is similar to the law of force which acts on a pendulum). There is still one point which requires explanation. If a body, or a lamina, is put in a state of vibration, frictions of all kinds will tend to reduce its vibrations; and it is necessary for us to shew that, under the actual circumstances, there is some force which tends to increase vibrations. Now Professor Willis suggested as probable that the magnitude of the force of air which acts on the lamina may be a function of the position of the lamina at a short time anterior to the time under consideration; and the author of this Treatise, in the *Cambridge Transactions,* Vol. IV., pointed out the mathematical consequences of that supposition. If the distance of a lamina from its place of rest be $\phi(t)$, and if e be the coefficient of the force dependent on laminar tension, k the coefficient of the force dependent on the action of the air, c the quantity to be subtracted from the time in order to form that time on which the force produced by the air-motion depends; then the equation of motion is,

$$\frac{d^2 . \phi(t)}{dt^2} = -e . \phi(t) - k . \phi(t-c).$$

Supposing k to be small, an approximate solution, obtained by neglecting the last term, is,

$$\phi(t) = a . \sin(t\sqrt{e} + b);$$

therefore

$$\phi(t-c) = a \,.\, \sin(t\sqrt{e} + b - c\sqrt{e}),$$

nearly; and

$$\frac{d^2 \,.\, \phi(t)}{dt^2} = -e \,.\, \phi(t) - ak \,.\, \sin(t\sqrt{e} + b - c\sqrt{e}),$$

very approximately.

This is precisely the case of a pendulum disturbed by some extraneous force; and it admits of being treated in the manner indicated by the author of this Treatise in a paper "On the disturbances of Pendulums, &c." in the *Cambridge Transactions*, Vol. III. The result is that, in every complete vibration, the coefficient of vibration is increased by $\frac{\pi ak \,.\, \sin c\sqrt{e}}{e}$.

104. *Experiments on the production of vowel sounds.*

Long ago, it had been shewn that if air were forced, through vibrating laminæ or through a common reed, into a cavity of a particular form, the sound of a particular vowel would be produced; but the forms of the cavities were very strange; and there was no theory, accounting for the effects in any case, or connecting the different cases. The great step of experimental explanation was made by Professor Willis in the *Cambridge*

Transactions, Vol. III.;* and nothing of importance has been added, till within a short time.

For instructive experiment, the strange forms of cavities were discarded, and a simple pipe was substituted; thus admitting of variation in its length. Air was blown through a long channel terminating in a reed that gave a note whose pitch was ascertained (and whose length of air-wave was therefore known); and the immediate carrier of this reed being fitted accurately but not tightly within the tube whose length was to be made variable, that tube was slid along so as to contain the reed-carrier as a plug in the tube, leaving open varying lengths of the tube between the reed-carrier and the external air. And the points remarked by Professor Willis are these:

First, that in order to perceive clearly a vowel-sound, it is necessary to sound different vowels in succession; the principal effect being produced by contrast, and no distinct vowel-sound being impressed on the ear when the apparatus is maintained steadily in the arrangement proper for producing any one vowel.

Second. The fundamental result of the experiments is this. In Figure 22, let *a* denote the place where the waves of air enter immediately from the reed (the reed being supposed to be at the left hand, and the current of air being blown through from left to right), and

* The author of this Treatise had the pleasure of *hearing* the original experiments.

suppose the tube of variable length to extend towards the right, its mouth sometimes stopping at I, sometimes being advanced to E, A, O, &c. Also let $ac = bd = ce$ = length of sound-wave produced by the reed. Then, when the mouth of the pipe is at the point I, it utters the vowel-sound I (in Continental pronunciation), the same vowel-sound as in the word "see"; when the mouth of the pipe is at E, the vowel-sound is that of E, sliding between the vowel-sounds in "pet" and "pay"; when the mouth of the pipe is at A, O, U, the sounds are respectively those in "paa, part", followed by "paw, nought"; in "no"; and in "but", followed by "boot". As the mouth of the pipe is carried towards b (the point bisecting ac), sound becomes indistinct, and vowel-sound is lost; or the only sound perceptible is that of our short U; on approaching c, the same vowels recur, but in the opposite order. On proceeding further still, the same phænomena recur after an addition to the pipe-length of "length of air-wave", "double length of air-wave", &c., but all sounds become less forcible.

Now upon varying the pitch of the reed (that is, upon varying the length of the sound-wave, or the length of the spaces ab, bc, cd, de), the lengths aI, aE, aA, aO, aU, cI, cE, &c., remain unaltered. And, when a reed of high pitch is used, or when the spaces ab, bc, &c., are made very short, some of the vowels U, O, &c., are lost. This accords with the experience of singers of high pitch, who can sing no vowel but I ("see").

The distances of the vowel-positions from c, or from e, are the following (the measure from a being a little smaller than the others):

$$
\begin{aligned}
& & \text{inch} \\
c\text{I} = e\text{I} &= & 0{\cdot}38 \\
c\text{E} = e\text{E} &= & \begin{cases} 0{\cdot}6 \\ 1{\cdot}0 \end{cases} \\
c\text{A} = e\text{A} &= & \begin{cases} 1{\cdot}8 \\ 3{\cdot}2 \end{cases} \\
c\text{A}' = e\text{A}' &= & \begin{cases} 3{\cdot}05 \\ 3{\cdot}8 \end{cases} \\
c\text{O} = e\text{O} &= & 4{\cdot}7 \\
c\text{U} \text{ or } e\text{U} & & \text{indefinite.}
\end{aligned}
$$

We recommend the reader to follow up this subject by a study of Professor Willis's original paper, and especially to remark his explanation, or rather his discrimination between the mechanical circumstances of the air in pipes of different lengths. It amounts to this; that from the air of invariable pressure at the pipe's mouth there is a species of reflexion of the sound-waves inwards (which is algebraically represented by the terminal equations for an open pipe,

$$-\phi'(at) + \psi'(at) = 0, \quad -\phi'(at - l) + \psi'(at + l) = 0,$$

in Article 79, implying opposite waves), and that thus every reed-wave travelling along the pipe is reflected from the open mouth at a time depending on the length of the pipe; the relation of which time to the time of the next reed-wave will be different for different lengths of the pipe, thus producing a mixed wave whose quality varies with the changes of that relation. (This amounts to nearly the same as saying that each puff of air through the reed may create a wave which will travel with organ-pipe-velocity, coexisting with one which follows the laws of Resonance; a coexistence which, as we have remarked in Article 81, is possible.) And the following experiment pointedly illustrates this. If a quill be snapped by the teeth of a wheel in rapid vibration, a musical note is produced; but if, instead of a quill, a highly elastic spring is used, itself competent to give a musical tone, then a vowel-sound is produced, and the name of the vowel depends on the relation between these two musical notes (which relation may be altered by grasping the elastic spring at different points).

The only addition, we believe, which has been made to this admirable series of experiments, is the further analysis, by Helmholz, of the character of the waves in the different vowel-sounds. It appears to have been made by directing the current of air from the lips upon a series of tuning-forks, and remarking which of the tuning-forks were put in vibration. Helmholz thus arrived at the conclusion that the vowel-sounds are produced by certain mixtures of the upper harmonics with the fundamental tone. It has been shewn in Article 72 that every variation of the quality of musical notes may be produced thus. But the distinct conclusion at which Professor Willis arrived, that, with different reed-notes, the difference in the linear measure of pipes required to produce different vowel-sounds must always be the same, seems to forbid the admission of *similar* mixtures of the upper harmonics with the fundamental tone as in all cases explaining the vowel-sounds.

The reader is referred to Willis in the *Cambridge Transactions,* Vol. III.; Wheatstone in the *London and Westminster Review,* October 1837—January 1838, page 27; Helmholz, *Die Lehre von den Tonempfindungen,* Part I. section 5, article 7; Tyndall, *Lectures,* page 200.

105. *Explanation of the formation of vowels and consonants by the human organs of speech.*

We are now in possession of all that is necessary to explain the vowels of the human voice. The musical note, which in experiments is produced by the reed, in nature is produced by the vocal ligaments of the glottis. For the varied length of pipe in Professor Willis's experiments, or for the conical or other receptacles of the wave opened to different degrees in the experiments of preceding and following experimenters, we have the cavity of the mouth and nostrils and communicating spaces, admitting of great changes at the will of the speaker. This appears to be all that is required.

On the production of consonants, we have little to remark. They depend on the mode of beginning or closing the utterance of the vowel-sounds. Sometimes this is done at the glottis, sometimes (in beginning the utterance) by lowering the tongue from the palate, sometimes by opening the lips, sometimes by opening the teeth. In some cases a vowel-sound must be formed before opening the lips; thus a momentary dull vowel-sound within the mouth, before opening the lips, appears necessary to give the sound *bee;* if there be no such antecedent dull vowel-sound, the sound emitted will be *pee.* (This appears to justify the Greek mediæval and modern writers, who to express the sound *b* in the languages of Western Europe use the combination *μπ*.) In closing the utterance of the vowel-sounds, there is nearly the same variety. All these different modifications give rise to different consonants; but they do not appear to involve any distinct principle which requires attention here. The roaring sound (that of *r*), the hissing sound (that of *s*), and the guttural sound, which however seems to be produced rather in the palate than in the throat (that of *ach* or *och*), with such dependencies as *sh, th,* appear to be abandonments of musical utterance, and probably do not require any action of the vocal ligaments; their peculiarities are given by the tongue, teeth, cheeks, and lips.

106. *Notes on the organs of hearing; deafness.*

Of the mechanism of hearing we know almost nothing. It is easily ascertained that the opening from the external air expands somewhat into a chamber where the wider part is closed by a stretched membrane called the *membrana tympani*, which in the healthy state is continuous, but which is sometimes punctured or slightly ruptured without producing great injury to the power of hearing. Beyond this is a cavern which communicates with the interior of the mouth by a passage called the Eustachian tube. The author of this Treatise, in descending to a great depth under water in a diving-bell, has suffered the pain (so often described) produced by the increase of atmospheric pressure on the exterior surface of the membrana tympani, which can be relieved at any depth by that motion of the swallowing muscles which enables a person to swallow the saliva; a motion that appears to open the Eustachian tube and to allow air to enter through it to the interior surface of the membrana tympani. But he also remarked that, in reascending, as the pressure of air within the mouth diminishes, the condensed air, next the interior surface of the membrana tympani, opens the Eustachian tube from time to time, and escapes into the mouth, giving the sensation of a loud crack.

The habitual state of the Eustachian tube being that of complete closure, the internal cavity resembles that of a kettle-drum; and there seems to be no doubt that vibration of the membrana tympani corresponding to the vibration of the air in the sounds which present themselves is—like the vibration of the parchment of the kettle-drum—the first element in the transmission of sound. Beyond this, we know nothing. There are small bones in a state of pressure, and there are curved tubes charged with fluid, and there are fibrous fringes; but we know not the function of any of them.

It seems likely that some advance might be made in our knowledge by a study of the phænomena of deafness. There is frequent misapprehension on the character of this loss of sense. In many cases, and perhaps ultimately in all cases, it is an incapability of hearing anything; but in a far greater number of cases it consists in hearing too much; in an excessive sensitiveness to certain sounds which, by a species of resonance, mix themselves with other sounds so as to produce confusion. In an instance that lately came under our observation, we remarked that the two notes which we have marked A♯ and a♯ produced a very painful effect, of the character of confused loud resonance. With this, or following this, was a frequent failure to perceive the hissing sound of *s*; and perhaps the general sensibility to sound diminished. But we are quite unable to say what part of the animal structure was the seat of the organic fault.

DESCRIPTION OF THE PLATES.

Figure 1 represents the state of the barometer-tube when, after the tube with its closed end downwards has been filled with quicksilver, and when a cup or cistern nearly filled with quicksilver has been prepared for its reception, the open end of the tube is covered (as with the finger), the tube is suddenly inverted, its open end is plunged deeply into the quicksilver of the cistern, and the finger or other covering is withdrawn. Then, if the length of the tube exceed 32 or 33 inches, the upper surface of the quicksilver drops, as shewn in the figure, leaving a vacuum above it, and exhibiting a column of quicksilver whose height, measured from the surface of the quicksilver in the cistern, represents the pressure of the atmosphere on that surface.

Figures 2 and 3 represent curved tubes in which the pressure of quicksilver (above which is vacuum) in a long column balances the aggregate of atmospheric pressure and the pressure of quicksilver in a short column.

Figures 4 and 5 represent curved tubes in which there is a certain quantity of air between the surface of the quicksilver and the closed end of the tube. In Figure 4, that air is allowed to expand itself, till its diminished elasticity, added to the pressure of a long column of quicksilver, balances the pressure of the atmosphere added to that of a short column of quicksilver. In Figure 5, the pressure of the atmosphere, added to that of a long column of quicksilver in the open tube, overcomes the pressure of a short column in the closed tube, and condenses the air above it till its elasticity is so much increased as to produce, with the weight of that short column, a balance of pressures.

Figure 6 represents the movements of particles of air in a horizontal column, when waves of condensation and expansion pass continuously through them from left to right. The first line represents the particles as at rest; the remaining five lines represent them in successive states as the waves advance successively more and more to the right. The delicate curved lines nearly vertical shew the successive positions of the same particle: the movement of every particle is obviously oscillatory. If we examine the motions of one particle, for instance g; we see that at the time T it was at its undisturbed or mean position, but moving backwards; but the condensation of particles before and behind were equal, so that there was no tendency to alter its velocity. But at the time $T+\frac{\tau}{4}$ it had gradually approached to and had reached a position in which, respect being also had to the positions gained by neighbouring particles, the condensation and consequent elastic force were much greater in rear than in front, and there was a tendency to throw it forward. Carrying on similar considerations to the times $T+\frac{2\tau}{4}$, $T+\frac{3\tau}{4}$, $T+\tau$, it will be seen that, at every time, the forces produced by the supposed motion of the particles are such as are able to maintain that supposed motion of the particles. It will also be seen that between the time T and the time $T + \tau$, every particle has described its complete double oscillation backward and forward; and also that the state of condensation has advanced in the same interval of times, from a to a', or through the interval between two waves.

In Figure 7, the motions of every particle (which are entirely vertical) may be imitated exactly by those of a stretched cord. The reader is referred for these to the author's Treatise *On the Undulatory Theory of Light.*

In Figure 8, the motions are more complicated, being compounded (for each particle) according to this law: that the horizontal parts of the motion are exactly those in Figure 6, and the vertical parts are exactly those in Figure 7: the place of greatest elevation corresponding to the place of greatest horizontal condensation, and the coefficients of vertical and horizontal displacement being sensibly equal. The author has treated of this subject in the *Encyclopœdia Metropolitana,* "Tides and Waves."

Figures 9, 10, 11, appear to require no explanation beyond that in the text.

In Figure 12, right-hand figure, is represented a crooked wire which can be pressed upon the elastic lamina or tongue of the Reed. By thrusting this to different positions, the vibrations of the reed-tongue may be made to correspond more nearly with those due to the length of the organ-pipe.

Figure 13 is explained in the text.

Figures 14 and 15 represent two complicated waves of the approximate concord of Octave formed with different intervals of the waves, which can in no wise be reconciled as similar, at whatever point either be supposed to commence; and thus the effect which their alternation produces in creating beats (see Article 92) fully justifies the opinion of practical musicians (see Article 87).

Figures 16 and 17 represent two complicated waves of the approximate concord of Fourth; which can be made a little more similar by beginning the first at its middle point, but which will always have a very sensible difference. They therefore produce beats (Article 92), but not so remarkable as those of the approximate Octave.

Figure 18 shews the intense beats produced by the approximate Unison.

In the second diagram of Figure 20 is exhibited a view of the top of the first diagram of Figure 20. Each of the laminæ is mounted as in Figure 19 (two laminæ facing each other), and the blast of air from below separates them, in concave forms.

In the second diagram of Figure 21 is exhibited a view of the top of the first diagram of Figure 21. When the blast of air enters from below between the two laminæ separated to a distance, it sucks them together with forms convex to each other.

Figure 22 is explained in the text.

ADDENDUM TO ARTICLE 54**.

It will be instructive to examine the law of radial movements of the particles of air, both without and within the spherical or hemispherical bell, as depending on the distances of the particles from the bell's center.

We have found that when $F = xyR$, while R is considered as a function of r and v only,

$$R = \frac{1}{3} \cdot \frac{d^2H}{dv^2} r^{-3} + \frac{dH}{dv} r^{-4} + Hr^{-5};$$

in which expression H is a function of v only, v being $= at - r$.

And, similarly, if we had considered R as a function of r and u only, u being $= at + r$, we should have found

$$R = \frac{1}{3} \cdot \frac{d^2L}{du^2} r^{-3} - \frac{dL}{du} r^{-4} + Lr^{-5},$$

where L is a function of u only.

And the most general expression for R is

$$R = \left\{ \begin{array}{l} \frac{1}{3} \cdot \frac{d^2H}{dv^2} r^{-3} + \frac{dH}{dv} r^{-4} + Hr^{-5} \\ + \frac{1}{3} \cdot \frac{d^2L}{du^2} r^{-3} - \frac{dL}{du} r^{-4} + Lr^{-5} \end{array} \right\}$$

T

Then the velocity of a particle of air in the direction of the radius from the center of the bell is

$$\frac{x}{r}\cdot\frac{dF}{dx}+\frac{y}{r}\cdot\frac{dF}{dy}+\frac{z}{r}\cdot\frac{dF}{dz},$$

or

$$xy\left\{\frac{2R}{r}+\frac{dR}{dr}\right\},$$

or

$$\frac{xy}{r^2}\cdot\frac{d\,(Rr^2)}{dr};$$

which becomes, in the general case,

$$\frac{xy}{r^2}\cdot\frac{d}{dr}\left\{\begin{array}{l}\frac{1}{3}\cdot\frac{d^2H}{dv^2}\,r^{-1}+\frac{dH}{dv}\,r^{-2}+Hr^{-3}\\ +\frac{1}{3}\cdot\frac{d^2L}{du^2}\,r^{-1}-\frac{dL}{du}\,r^{-2}+Lr^{-3}\end{array}\right\}.$$

It will be remarked here that $\frac{xy}{r^2}$ is purely a function of the angle of inclination of the radius in question to x, y, and z; not changing its value while the direction of the radius remains the same, and never becoming indefinitely great.

Performing the actual differentiation of the expression above, and remembering that

$$\frac{dv}{dr}=-1,\ \ \frac{du}{dr}=+1,$$

this becomes

$$\frac{xy}{r^2}\left\{\begin{array}{l}-\frac{1}{3}\cdot\frac{d^3H}{dv^3}\,r^{-1}-\frac{4}{3}\cdot\frac{d^2H}{dv^2}\cdot r^{-2}-3\cdot\frac{dH}{dv}\,r^{-3}-3Hr^{-4}\\ +\frac{1}{3}\cdot\frac{d^3L}{du^3}\,r^{-1}-\frac{4}{3}\cdot\frac{d^2L}{du^2}\cdot r^{-2}+3\cdot\frac{dL}{du}\,r^{-3}-3Lr^{-4}\end{array}\right\}.$$

First, we will consider the radial movement of the particles of air external to the bell.

Let b be the radius of the bell. The radial movement of the surface of the bell (which must be the same as the radial movement of the particles of air in contact with the bell, external and internal) will be found by putting b for r, in the functions

$$H, L, \frac{dH}{dv}, \frac{dL}{du}, \text{\&c.}$$

as well as in the explicit factors. And, as the movement of the bell is arbitrary, the forms of the functions must be adapted to represent that arbitrary motion when $r = b$. It will be remembered that all functions of v or $at - r$ then become functions of $at - b$, and that all functions of u or $at + r$ become functions of $at + b$, in which the only variable is t.

Algebraically, we may take any forms for H and L which upon substituting b for r will cause the function, that we have lately found, to represent the movement given by arbitrary causes to the bell. No terms become infinite, and there is no other algebraical condition whatever. But practically there is this condition; we suppose the waves to begin at the bell and to flow outwards; and therefore we must reject the terms depending on u, and retain those depending on v. In any simple assumption for the motion of the bell, as for instance $\frac{xy}{r^2} A . \sin (Bt + C)$, (or any assemblage of similar terms with different constants,) there will be no difficulty in finding the general form of the function H which, when it is substituted in the bracket above, and when finally b is put for r, gives the assumed value for the motion of the bell.

Secondly, in regard to the movement of the particles of air within the bell.

It is impossible to define these by means of the functions of v only (contained in the first line of the bracket), because when $r=0$ they receive infinite multipliers, and there is no tendency in the different terms to destroy each other. The only possibility of giving a solution which is admissible depends on the combination of functions of v and u. We will examine into the practicability of so arranging these that no infinite terms shall remain.

We will first make a change which is convenient, though not necessary, in the expression for R.

$$\frac{d}{dr}(Lr^{-1}) = \frac{dL}{du}r^{-1} - Lr^{-2};$$

and $$r^{-1}.\frac{d}{dr}(Lr^{-1}) = \frac{dL}{du}r^{-2} - Lr^{-3}.$$

Differentiating this,

$$\frac{d}{dr}\left\{r^{-1}.\frac{d}{dr}(Lr^{-1})\right\} = \frac{d^2L}{du^2}r^{-2} - 3\frac{dL}{du}r^{-3} + 3Lr^{-4};$$

and $\frac{1}{3} r \cdot \frac{d}{dr}\left\{r^{-1} \cdot \frac{d}{dr}(Lr^{-1})\right\} = \frac{1}{3} \cdot \frac{d^2L}{du^2} r^{-1} - \frac{dL}{du} r^{-2} + Lr^{-3}.$

This is the same as the term in R which depends on L. And therefore the term in the radial motion of the air which depends upon L may be expressed by

$$\frac{1}{3} \cdot \frac{xy}{r^2} \cdot \frac{d}{dr}\left[r \cdot \frac{d}{dr}\left\{r^{-1} \cdot \frac{d}{dr}(Lr^{-1})\right\}\right].$$

And it will be found on trial, giving due attention to the circumstance that $\frac{dH}{dr} = -\frac{dH}{dv}$, that the very same expression applies to the term depending upon H, only substituting H for L.

For the air within the bell, let M and N be the functions of v and u. It is at present our special object to examine the movements when r is very small or 0; and therefore we will expand the expressions in powers of r. Let M_0, ${}^1M_0$, ${}^2M_0$, ${}^3M_0$, ${}^4M_0$, ${}^5M_0$, &c., be the values of M, $\frac{dM}{dv}$, $\frac{d^2M}{dv^2}$, $\frac{d^3M}{dv^3}$, &c., when $r=0$; or when v, which $= at - r$, is reduced to at. These quantities M_0, ${}^1M_0$, &c., are functions of t, but not of r. And let N_0, ${}^1N_0$, &c., be quantities having the same relation to N. Then the radial displacement of the air,

or $$\frac{1}{3} \frac{xy}{r^2} \cdot \frac{d}{dr}\left[r \cdot \frac{d}{dr}\left\{r^{-1} \cdot \frac{d}{dr}(\overline{M+N} \cdot r^{-1})\right\}\right],$$

will be found by substituting the following values for M and N;

$$M = M_0 - {}^1M_0 . r + {}^2M_0 . \frac{r^2}{2} - {}^3M_0 \frac{r^3}{2.3} + {}^4M_0 \frac{r^4}{2.3.4} - {}^5M_0 \frac{r^5}{2.3.4.5} + \&c.$$

$$N = N_0 + {}^1N_0 . r + {}^2N_0 . \frac{r^2}{2} + {}^3N_0 \frac{r^3}{2.3} + {}^4N_0 . \frac{r^4}{2.3.4} + {}^5N_0 . \frac{r^5}{2.3.4.5} + \&c.$$

the general term of $M + N$ being, for every term except the first,

$$\{(-1)^n . {}^nM_0 + {}^nN_0\} \times \frac{r^n}{1.2.3...n}.$$

The corresponding general term of radial displacement, formed by use of the expression above, will be

$$\frac{1}{3} . \frac{xy}{r^2} . \{(-1)^n . {}^nM_0 + {}^nN_0\} \times \frac{(n-1).(n-3)^2}{1.2.3......n} r^n;$$

and, forming the first term by independent use of the expression, and forming the other terms by making n successively $= 1, 2, 3,$ &c., we obtain for the radial displacement,

$$\frac{1}{3} \frac{xy}{r^2} \times \left\{ \begin{array}{l} -(M_0 + N_0) . 9r^{-4} + ({}^2M_0 + {}^2N_0) \frac{r^{-2}}{2} + ({}^4M_0 + {}^4N_0) . \frac{1}{2.4} \\ + (-{}^5M_0 + {}^5N_0) . \frac{2r}{3.5} + \&c. \end{array} \right\}.$$

Now, if $M = \psi(v)$, let $N = -\psi(u)$. Then, when $r = 0$, $M = \psi(at)$, $N = -\psi(at)$, and $M_0 + N_0 = 0$.

The same holds for the sum of every pair of differential coefficients of even order: those of odd orders have equal values with similar signs. And the entire expression for radial motion is

$$\frac{4}{45} \cdot \frac{xy}{r^2} \cdot {}^5N_0 \cdot r + \&c.,$$

and there is no infinite term.

Thus the only condition which, algebraically, requires attention, is that the functions of u and v be similar in form, but affected with opposite signs.

The radial movement, derived from an expression above, with attention to this consideration, will now be

$$\frac{xy}{r^2} \times \left\{ \begin{array}{l} -\frac{1}{3} \cdot \frac{d^3 . \psi(v)}{dv^3} r^{-1} - \frac{4}{3} \cdot \frac{d^2 . \psi(v)}{dv^2} . r^{-2} - 3 . \frac{d . \psi(v)}{dv} r^{-3} - 3 . \psi(v) r^{-4} \\ -\frac{1}{3} \cdot \frac{d^3 \psi(u)}{du^3} r^{-1} + \frac{4}{3} \cdot \frac{d^2 . \psi(u)}{du^2} . r^{-2} - 3 . \frac{d . \psi(u)}{du} r^{-3} + 3 . \psi(u) r^{-4} \end{array} \right\}.$$

The form of the function ψ is to be determined by trial so that, when the several differentiations are performed and b is substituted for r, the radial movement will correspond with the assumed motion of the bell. If the law of assumed motion of the bell be simple, as for instance $\frac{xy}{r^2} A . \sin(Bt + C)$, (or any assemblage of similar terms with different constants,) there will be no difficulty in finding a form for ψ.

ADDENDUM TO ARTICLE 61.

A series of experiments on the properties of Steel, including amongst other things the rates of extensibility and compressibility by the action of forces, has lately been published by the Institution of Civil Engineers. It appears from these that the amount of compression produced by a given weight acting for compression, and the amount of extension produced by the same weight acting for extension, are sensibly the same. And that, for a bar whose section is 1 square inch, a force of 1 ton employed in extension or compression alters the length of the bar by 0·000076 × the whole length of the bar. The specific gravity of the steel was 7·847.

As the weight of a cubic foot of water is (with sufficient accuracy for these experiments) 1000 ounces avoirdupois, the weight of a cubic foot or 1728 cubic inches of steel is about 7847 ounces or 490·31 pounds; and the number of cubic inches of steel (or the length in inches of a steel bar of 1 inch section) weighing 1 ton or 2240 pounds will be 7895. As this produces an extension or compression of 0·000076 of the whole length, the constant L in the expressions above will be in inches $\frac{7895}{0·000076} = 103882000$; or in feet $= 8657000$. From this, by the formula above, the velocity of sound through steel $= 16672$ feet per second.